Introduction to
Graphics Communications
for Engineers

Welcome to the BEST!

McGraw-Hill's *BEST*—**B**asic **E**ngineering **S**eries and **T**ools—consists of modularized textbooks and applications appropriate for the topic covered in mostly introductory engineering courses. The goal of the series is to provide the educational community with material that is timely, affordable, of high quality, and flexible in how it is used.

McGraw-Hill's BEST!

Introduction to
Graphics Communications
for Engineers

Second Edition

Gary R. Bertoline
Purdue University

McGraw Hill

Boston   Burr Ridge, IL   Dubuque, IA   Madison, WI   New York   San Francisco   St. Louis
Bangkok   Bogotá   Caracas   Kuala Lumpur   Lisbon   London   Madrid   Mexico City
Milan   Montreal   New Delhi   Santiago   Seoul   Singapore   Sydney   Taipei   Toronto

McGraw-Hill Higher Education

A Division of The McGraw-Hill Companies

INTRODUCTION TO GRAPHICS COMMUNICATIONS FOR ENGINEERS
SECOND EDITION

Published by McGraw-Hill, a business unit of The McGraw-Hill Companies, Inc., 1221
Avenue of the Americas, New York, NY 10020. Copyright © 2002, 1999 by The McGraw-Hill
Companies, Inc. All rights reserved. No part of this publication may be reproduced or distributed
in any form or by any means, or stored in a database or retrieval system, without the prior written
consent of The McGraw-Hill Companies, Inc., including, but not limited to, in any network or
other electronic storage or transmission, or broadcast for distance learning.

Some ancillaries, including electronic and print components, may not be available to customers
outside the United States.

This book is printed on recycled, acid-free paper containing 10% postconsumer waste.

International   1 2 3 4 5 6 7 8 9 0 QPD/QPD 0 9 8 7 6 5 4 3 2 1
Domestic   1 2 3 4 5 6 7 8 9 0 QPD/QPD 0 9 8 7 6 5 4 3 2 1

ISBN 0–07–243634–4
ISBN 0–07–112171–4 (ISE)

General manager: *Thomas E. Casson*
Publisher: *Elizabeth A. Jones*
Editorial coordinator: *Zuzanna Borciuch*
Marketing manager: *Ann Caven*
Project manager: *Mary E. Powers*
Production supervisor: *Kara Kudronowicz*
Coordinator of freelance design: *David W. Hash*
Photo research coordinator: *John C. Leland*
Photo research: *Randall Nicholas*
Compositor: *Precision Graphics*
Typeface: *10/12 Times Roman*
Printer: *Quebecor World Dubuque, IA*
Image on Chapter Opener: *PhotoDisc/AA 016 423/Vol. 85*

INTERNATIONAL EDITION ISBN 0–07–112171–4
Copyright © 2002. Exclusive rights by The McGraw-Hill Companies, Inc., for manufacture and
export. This book cannot be re-exported from the country to which it is sold by McGraw-Hill.
The International Edition is not available in North America.

www.mhhe.com

For Ada, Bryan, Kevin, and Carolyn, who are my
motivation and inspiration for all my books.

Gary R. Bertoline

Gary R. Bertoline is Professor and Department Head of Computer Graphics Technology at Purdue University. He earned his B.S. degree in Industrial Technology at Northern Michigan University in 1974, M.Ed. in Industrial Technology at Miami University in 1979, and Ph.D. at The Ohio State University in Industrial Technology in 1987. His graduate work focused on the integration of CAD into engineering graphics and visualization. He has 25 years' experience teaching graphics at all levels from elementary school to senior citizens. Prof. Bertoline taught junior high and high school graphics at St. Henry High School, St. Henry, Ohio; drafting/design technology at Wright State University, Lake Campus, Celina, Ohio; and engineering graphics at The Ohio State University, Columbus, Ohio.

Prof. Bertoline has authored numerous publications, authored or co-authored 10 textbooks and workbooks, and made over 50 presentations throughout the world. He has won the Frank Oppenheimer Award three times for best paper at the Engineering Design Graphics Division Mid-year Meeting. He has developed many graphics courses, including CAD, solid modeling, and multimedia, and has integrated many modern topics into traditional engineering graphics courses, such as modeling, animation, and visualization. Prof. Bertoline has conducted research in cognitive visualization and was the co-author for a curriculum study in engineering graphics funded by SIGGRAPH. He is on the editorial board for the *Journal for Geometry and Graphics* and is the McGraw-Hill Graphics Series Editor. He was the recipient of the Orthogonal Medal for outstanding contributions to the advancement of Graphic Science by North Carolina State University in 1992, and the 1995 inaugural recipient of the Steve M. Slaby International Award for Outstanding Contributions in Graphics Education. You can contact Dr. Bertoline at grbertol@tech.purdue.edu.

Contents

Foreword

Engineering educators have had long-standing debates over the content of introductory freshman engineering courses. Some schools emphasize computer-based instruction, some focus on engineering analysis, some concentrate on graphics and visualization, while others emphasize hands-on design. Two things, however, appear certain: No two schools do exactly the same thing, and at most schools, the introductory engineering courses frequently change from one year to the next. In fact, the introductory engineering courses at many schools have become a smorgasbord of different topics, some classical and others closely tied to computer software applications. Given this diversity in content and purpose, the task of providing appropriate text material becomes problematic since every instructor requires something different.

McGraw-Hill has responded to this challenge by creating a series of modularized textbooks for the topics covered in most first-year introductory engineering courses. Written by authors who are acknowledged authorities in their respective fields, the individual modules vary in length, in accordance with the time typically devoted to each subject. For example, modules on programming languages are written as introductory-level textbooks, providing material for an entire semester of study, whereas modules that cover shorter topics such as ethics and technical writing provide less material, as appropriate for a few weeks of instruction. Individual instructors can easily combine these modules to conform to their particular courses. Most modules include numerous problems and/or projects and are suitable for use within an active-learning environment.

The goal of this series is to provide the educational community with text material that is timely, affordable, of high quality, and flexible in how it is used. We ask that you assist us in fulfilling this goal by letting us know how well we are serving your needs. We are particularly interested in knowing what, in your opinion, we have done well, and where we can make improvements or offer new modules.

Byron S. Gottfried
Consulting Editor
University of Pittsburgh

Preface

Introduction to Graphics Communications for Engineers, 2nd Edition is part of the McGraw-Hill's BEST (Basic Engineering Series and Tools), which introduces engineering students to various topics important to their education. This workbook is an introduction to the standard practices used by engineers to communicate graphically. The primary goal of this text is to assist engineering students in learning the techniques and standards of communicating graphically so that design ideas can be clearly communicated and produced.

The text concentrates on the concepts and skills necessary to sketch and create 2-D drawings and 3-D CAD models. Engineering graphics has gone through significant changes in the last decade as a result of the use of computers and CAD software. It seems as if some new hardware or software development that impacts engineering graphics occurs every year. Although these changes are important to the subject of technical graphics, there is much about the subject that has not changed. Engineers still find it necessary to communicate and interpret design ideas through the use of graphics methods such as sketches and CAD drawings and models. As powerful as today's computers and CAD software have become, they are of little use to engineers who do not fully understand fundamental graphics communications principles and 3-D modeling strategies, or who do not possess a high level of visualization ability.

The workbook is divided into six chapters with multiple units of instruction. Chapter 1, "Introduction to Graphics Communications," is an introduction to graphics communications as a language for engineers and describes the tools used and some of the techniques for communicating graphically. Chapter 2, "Sketching and Text," is an introduction to sketching technique, projection theory, visualization, and the use of text on drawings. Chapter 3,

"Section and Auxiliary Views," introduces the student to the use of and technique for creating sectioned drawings and models and auxiliary views. Chapter 4, "Dimensioning and Tolerancing Practices," describes how to create and read dimensional drawings. Chapter 5, "Reading and Constructing Working Drawings," describes how to read and produce working drawings. Finally, Chapter 6, "Design and 3-D Modeling," is an overview of 3-D modeling techniques and the engineering design process.

Second Edition improvements include additional problems, a new design problem, some additional topics as requested by 1st edition users, and additional sketching techniques in Chapter 1.

Each chapter has a list of objectives, step-by-step instructions, and a wide assortment of problems that can be assigned to reinforce the topics covered. Some chapters have a list of further readings, so that students can find more information related to the topics covered in the chapter. Sketching worksheets have been integrated into the end of each chapter. These worksheets can be used for sketching assignments to augment assignments using CAD. After completing the workbook, the student will be able to create design sketches using various projection techniques, create and read 2-D standard engineering drawings, and create and visualize 3-D computer models.

Special thanks to James Mohler and Amy Fleck for their work on the illustrations and Jim Leach for some of the drawing problems added in the second edition.

Gary R. Bertoline, PhD
Professor and Department Head of
Applied Computer Graphics
Purdue University
West Lafayette, IN

Chapter

Introduction to Graphics Communication

1

OBJECTIVES

After completing this chapter, you will be able to:

1. Describe why technical drawings are an effective communications system for technical ideas about designs and products.
2. Identify important parts of a CAD system.
3. Identify important traditional tools.
4. Identify standard metric and U.S. drawing sheet sizes.
5. Identify the types and thicknesses of the various lines in the alphabet of lines.

1.1 | INTRODUCTION

Graphics communications using engineering drawings and models is a language—a clear, precise language—with definite rules that must be mastered if you are to be successful in engineering design. Once you know the language of graphics communications, it will influence the way you think, the way you approach problems. Why? Because humans tend to think using the languages they know. Thinking in the language of technical graphics, you will visualize problems more clearly and will use graphic images to find solutions with greater ease.

In engineering, 92 percent of the design process is graphically based. The other 8 percent is divided between mathematics and written and verbal communications. Why? Because graphics serves as the primary means of communication for the design process. Figure 1.1 shows a breakdown of how engineers spend their time. 3-D Modeling and documentation, along with design modeling, comprise more than 50 percent of the engineer's time and are purely visual and graphical activities. Engineering analysis depends largely on reading technical graphics, and

25% ⬛ 3-D Modeling and Documentation
19% ⬛ Manufacturing Engineering
25% ⬛ Functional Design
15% ⬛ Engineering Analysis
16% ☐ Other

Figure 1.1   A Total View of Engineering Divided into Its Major Activities
Graphics plays a very important role in all areas of engineering, for documentation, communications, design, analysis, and modeling. Each of the activities listed is so heavily slanted toward graphics communications that engineering is 92 percent graphically based. Information from Dataquest, Inc. CAD/CAM/CAE/GIS Industry Service.

manufacturing engineering and functional design also require the production and reading of graphics.

Why do graphics come into every phase of the engineer's job? To illustrate, look at the jet aircraft in Figure 1.2. Like any new product, it was designed for a specific task and within specified parameters; however, before it could be manufactured, a 3-D model and engineering drawings like that shown in Figure 1.3 had to be produced. Just imagine trying to communicate all the necessary details verbally or in writing. It would be impossible!

A designer has to think about the many features of an object that cannot be communicated with verbal descriptions. (Figure 1.4) These thoughts are dealt with in the mind of the designer using a visual, nonverbal process. This "visual image in the mind" can be reviewed and modified to test different solutions before it is ever communicated to someone else. As the designer draws a line on paper or creates a solid cylinder image with a computer, he or she is translating the mental picture into a drawing or model that will produce a similar picture in the mind of anyone who sees the drawing. This drawing or graphic representation is the medium through which visual images in the mind of the designer are converted into the real object.

Technical graphics can also communicate solutions to technical problems. Such technical graphics are produced according to certain standards and conventions so they can be read and accurately interpreted by anyone who has learned those standards and conventions.

Figure 1.2
This jet aircraft would be impossible to create without computer graphics models and drawings. Drawings are the road maps that show how to manufacture or build products and structures.

The precision of technical graphics is aided by tools; some are thousands of years old and still in use today, and others are as new and rapidly changing as computer-aided design/drafting (CAD). This book will introduce you to the standards, conventions, techniques, and tools of technical graphics and will help you develop your technical skills so that your design ideas become a reality.

Engineers are creative people who use technical means to solve problems. They design products, systems, devices, and structures to improve our living conditions. Although problem solutions begin with thoughts or images in the mind of the designer, presentation devices and computer graphics hardware and software are powerful tools for communicating those images to others. They can also aid the visualization process in the mind of the designer. As computer graphics have a greater impact in the field of engineering, engineers will need an ever-growing understanding of and facility in graphics communications.

Practice Exercise 1.1

1. Attempt to describe the part shown in Figure 1.10 using written instructions. The instructions must be of such detail that another person can make a sketch of the part.

2. Now try verbally describing the part to another person. Have the person make a sketch from your instructions.

These two examples will help you appreciate the difficulty in trying to use written or verbal means to describe even simple mechanical parts. Refer to Figure 1.3 and others in this text to get an idea of how complicated some parts are compared

Figure 1.3   Engineering Drawing
Engineering drawings and computer models such as these were needed to produce an aircraft. The 3-D model is used to design and visualize the aircraft. The engineering drawings are used to communicate and document the design process. Courtesy of the Northrop Grumman Corporation.

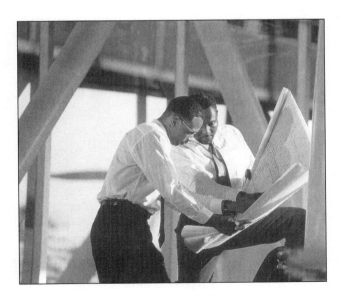

Figure 1.4   Engineering Drawings Used for Communications
Engineering drawings are a nonverbal method of communicating information. Descriptions of complex products or structures must be communicated with drawings. A designer uses a visual, nonverbal process. A visual image is formed in the mind, reviewed, modified, and is ultimately communicated to someone else, all using visual and graphics processes.

with this example. It is also important to note that air and water craft have thousands of parts. For example, the nuclear powered *Sea Wolf* class submarine has more than two million parts. Try using verbal or written instructions to describe that!

1.2 | WHAT YOU WILL LEARN

In this text, you will learn the six important areas in engineering graphics:

Visualization—the ability to mentally control visual information.
Graphics theory—geometry and projection techniques.
Standards—sets of rules that govern how parts are made and technical drawings are represented.
Conventions—commonly accepted practices and methods used for technical drawings.
Tools—devices used to create engineering drawings and models, including both hand-held and computer tools.
Applications—the various uses for technical graphics in engineering design, such as mechanical, electrical, and architectural.

Each chapter in the text will explain the graphics theory important for a topic, integrate the visualization practices, explain the relevant standards and conventions,

Design Case Study
3-D Design Pays Off for Bose Corporation

The challenge facing Bose® engineers was to fit an AM/FM tuner and six-disk CD player into an area that previously held only a single-disk CD player. Complicating the effort was that the final product was to be the smallest offering available on the market, yet needed to contain Bose's renowned styling and functionality. Time constraints mandated that the product be brought to market as quickly as possible. Objectives included:

- Engineer the product such that the access door to the CD magazine was on the front of the console. Additionally, the door had to function smoothly and be a high-quality component.

- Fit the CD changer, AM/FM tuner, and all the electronics needed to drive the changer, tuner, and display into an enclosure measuring 15.5 inches long by 8 inches wide by 2.5 inches high.

- Bring the Bose Lifestyle® 20 music system to market in record time.

- Demonstrate the effectiveness of its new 3-D software system to speed the product development cycle.

- The number one criterion for the new software: reducing time to market.

- Maintain the elegant design of Bose audio equipment, with a minimum of buttons and knobs cluttering the design.

The process began when Bose® engineers modeled each of the individual console components. The next step was to pull all the components into a master model of the assembly, so engineers could see how everything fit together. Then, the design team studied the fit of the components and assemblies within the enclosure. Shaded solid models made it easy to visualize the box and its many contents, and in many cases, this was all that was needed to spot problems and optimize the internal configuration. Designing the console door was next using 3-D CAD software to let the team examine the motion of the door directly from the solid model assembly, without creating special mechanism models.

Results
- Bose brought a superior product to market in record time.
- Bose Lifestyle® 20 music system was the smallest multiple-disk CD player on the market.
- The final product contained the sleek, elegant design of all Bose audio equipment offerings.
- The use of 3-D solid models made it much easier for the design team to communicate with electronics designers, manufacturing engineers, and vendors on the project. They conveyed design intent earlier and more accurately than previously possible. This enhanced communications reduced errors in downstream operations and also made it possible to perform analysis, get quotes, and make tools sooner because these processes didn't have to wait for drawings.

(Courtesy of Bose Corporation.)

demonstrate the tools used to create drawings, and apply the topic to engineering design.

1.3 | TECHNICAL DRAWING TOOLS

Just as the graphics language has evolved over the years into a sophisticated set of standards and conventions, so have the tools used to graphically communicate technical ideas. Tools are used to produce three basic types of drawings: freehand sketches, instrument drawings, and computer drawings and models. The tools have evolved from pencils, triangles, scales, and compasses to **computer-aided design/drafting (CAD)** systems. **CAD** is computer software and related computer hardware that supplements or replaces traditional hand tools in creating models and technical drawings. (Figure 1.5)

Figure 1.5 CAD Workstations
Typical CAD workstations used in industry have large color monitors.

Figure 1.6 Traditional Tools
These are some of the many traditional mechanical drawing tools used for engineering drawings.

Since many industries have not fully integrated CAD into their design offices, it is necessary to learn both traditional and computer design methods. Also, traditional tools are used for sketching, which is one of the most effective methods available to represent design ideas quickly.

1.4 | COMPUTER-AIDED DRAWING TOOLS

Traditional tools will continue to be useful for sketching and rough layout work; however, good CAD software can create virtually any type of technical drawing. Circle commands replace the compass, line commands replace the T-square and triangles, and editing commands replace the dividers and erasing shield.

A CAD system consists of hardware devices used in combination with specific software. The **hardware** for a CAD system consists of the physical devices used to support the CAD software. There are many different hardware manufacturers and types of hardware devices, all of which are used to create, store, or output technical drawings and models. It is not uncommon in industry to have multiple input, storage, and output devices for a CAD system.

1.5 | TRADITIONAL TOOLS

The traditional tools used to create technical drawings have evolved over time. Many tools were originally used in ancient Greece to study and develop geometry. Al-

though computers may someday replace the need for some traditional tools, they are still useful today for drawing, and more importantly, for sketching. **Traditional tools** are devices used to assist the human hand in making technical drawings. The assistance includes drawing lines straighter, making circles more circular, and increasing the speed with which drawings are made. The tools typically used to create mechanical drawings or sketches (Figure 1.6) consist of the following:

1. Wood and mechanical pencils.
2. Instrument set, including compass and dividers.
3. 45- and 30/60-degree triangles.
4. Scales.
5. Irregular curves.
6. Protractors.
7. Erasers and erasing shields.
8. Drawing paper.
9. Circle templates.
10. Isometric templates.

1.6 | MEDIA

Media are the surfaces upon which an engineer or technologist communicates graphical information. The media used for technical drawings are different types or grades of paper, such as tracing paper, vellum, and polyester film. Tracing paper is a thin, translucent paper used for detail drawings. Vellum is a tracing paper chemically treated to improve translucency. Polyester film, or its trade name Mylar, is transparent, waterproof, and difficult to tear.

Table 1.1 ANSI Standard Sheet Sizes

Metric (mm)	U.S. Standard	Architectural
A4 210 × 297	A-Size 8.5″ × 11″	9″ × 12″
A3 297 × 412	B-Size 11″ × 17″	12″ × 18″
A2 420 × 524	C-Size 17″ × 22″	18″ × 24″
A1 594 × 841	D-Size 22″ × 34″	24″ × 36″
A0 841 × 1189	E-Size 34″ × 44″	36″ × 48″

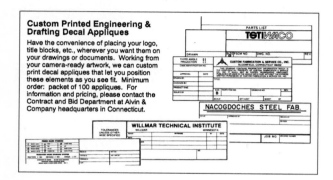

Figure 1.7 **Preprinted Title Blocks**
Preprinted standard borders and title blocks on drafting paper are commonly used in industry. Courtesy of Alvin & Company.

Mylar can be used for lead pencil, plastic-lead pencil, or ink drawings. Mylar is an excellent drawing surface that leaves no trace of erasure.

Special papers have also been developed for CAD plotters. For example, plotter paper used for fiber-tipped pens has a smooth or glossy surface to enhance line definition and minimize skipping. Often, the paper comes with a preprinted border, title block, and parts list. (Figure 1.7)

The American National Standards Institute (ANSI) has established standard sheet sizes and title blocks for the media used for technical drawings. Each paper size is designated by a letter, as shown in Table 1.1, and title block sizes are shown in Figure 1.11 at the end of the chapter.

1.7 | ALPHABET OF LINES

The **alphabet of lines** is a set of standard linetypes established by the American Society of Mechanical Engineers (ASME) for technical drawing. Figure 1.8 shows the alphabet of lines, and the approximate dimensions used to create different linetypes, which are referred to as

linestyles when used with CAD. ASME Y14.2M-1992 has established these linetypes as the standard for technical drawings. Two line weights are sufficient to follow the standards, a 0.6 mm and a 0.3 mm. These approximate widths are intended to differentiate between thin and thick lines and are not for control of acceptance or rejection of drawings. Thick lines are drawn using soft lead, such as F or HB. Thin lines are drawn using a harder lead, such as H or 2H. Construction lines are very light and are drawn using 4H or 6H lead. A good rule of thumb for creating construction lines is to draw them so that they are difficult to see if your drawing is held at arm's length.

Listed below are the standard linetypes and their applications in technical drawings:

Center lines are used to represent symmetry and paths of motion and to mark the centers of circles and the axes of symmetrical parts, such as cylinders and bolts.
Break lines come in two forms: a freehand thick line and a long, ruled thin line with zigzags. Break lines are used to show where an object is broken to save drawing space or reveal interior features.
Dimension and extension lines are used to indicate the sizes of features on a drawing.
Section lines are used in section views to represent surfaces of an object cut by a cutting plane.
Cutting plane lines are used in section drawings to show the locations of cutting planes.
Visible lines are used to represent features that can be seen in the current view.
Hidden lines are used to represent features that cannot be seen in the current view.
Phantom lines are used to represent a movable feature in its different positions.
Stitch lines are used to indicate a sewing or stitching process.
Chain lines are used to indicate that a surface is to receive additional treatment.
Symmetry lines are used as an axis of symmetry for a particular view.

It is important that you understand and remember these different linetypes and their definitions and uses, because they are referred to routinely throughout the rest of this book.

CAD software provides different linestyles for creating standard technical drawings. Figure 1.9 shows the linestyle menu for a typical CAD system. The thicknesses of lines on a CAD drawing are controlled by two different means: (1) controlling the thickness of the lines drawn on

Figure 1.8   The Alphabet of Lines

The alphabet of lines is a set of ASME standard linetypes used on technical drawings. The approximate dimensions shown on some linetypes are used as guides for drawing them with traditional tools. The technical drawing at the top shows how different linetypes are used in a drawing.

Figure 1.9   AutoCAD's Linestyle Menu Showing Some of the Linetypes Available

the display screen and (2) controlling the plotted output of lines on pen plotters by using different pen numbers for different linestyles, where different pen numbers have different thicknesses, such as a 0.7 mm and 0.3 mm.

Further Reading

Booker, P. *A History of Engineering Drawing.* London: Chatto & Windus, 1962.

Ferguson, E. S. "The Mind's Eye: Nonverbal Thought in Technology." *Science* 197, no. 4306 (August 26, 1977), pp. 827–36.

Grown, J. R., et al. *Visualization: Using Computer Graphics to Explore Data and Present Information.* New York: John Wiley and Sons.

Higbee, F. G. "The Development of Graphical Representations." In *Proceedings of the Summer School for Drawing Teachers.* Eds. R. P. Hoelscher, J. Rising. New York: McGraw Hill, 1949, pp. 9–26.

LaCourse, D. E. *Handbook of Solid Modeling.* New York: McGraw-Hill, 1995.

Land, M. H. "Historical Developments of Graphics." *Engineering Design Graphics Journal* 40, no. 2 (Spring 1976), pp. 28–33.

Reynolds, T. S. "Gaspard Monge and the Origins of Descriptive Geometry." *Engineering Design Graphics Journal* 40, no. 2 (Spring 1976), pp. 14–19.

Problems

1.1 Research and report on an important historical figure in engineering design, such as Henry Ford, Thomas Edison, the Wright brothers, or Alexander Graham Bell.

1.2 Identify at least five other individuals who worked as engineers and had an impact on society.

1.3 Research and report on an important historical engineering achievement, such as airplanes, space flight, computers, or television.

1.4 Identify three new products that have appeared on the market in the last five years.

1.5 Research and report on an important historical figure in graphics, such as Gaspard Monge, M. C. Escher, Thomas Alva Edison, Leonardo da Vinci, Albrecht Durer, or Frank Lloyd Wright.

1.6 To demonstrate the effectiveness of graphics communications, write a description of the object shown in Figure 1.10. Test your written description by having someone attempt to make a sketch from your description.

1.7 Make a sketch of a common device, such as a telephone, automobile, computer mouse, or coffee cup.

1.8 Get a clear mental picture of a television, then sketch what you see in your mind. Is this mental image 2-D or 3-D? Try to put words to each feature of the TV you are drawing. In this problem you will experience the difficulty in trying to verbally describe an object with enough detail for it to be manufactured.

1.9 Interview a practicing engineer or technologist and ask how graphics are used in his or her daily work.

1.10 Ask the practicing engineer or technologist what changes are taking place in his or her profession.

1.11 Research and report on an important historical figure in computer graphics, such as Ivan Sutherland, Steve Coons, R. E. Bezier, or George Lucas.

1.12 Draw the border lines and title blocks for the ANSI and ISO drawing sheets, using the dimensions shown. Add text as shown, using ⅛″ (3-mm) all caps text. (Figure 1.11)

Figure 1.10   Problem 1.6   Bearing Block to Be Described Verbally

**Figure 1.11   Problem 1.12
ANSI Standard Title Blocks
and Border Lines**

Title Block for A, B, C.

Title Block for D and E.

Continuation Sheet Title Block for A, B, C.

Continuation Sheet Title Block
for D and E.

Size Designation	Width (Vertical)	Length (Horizontal)	Margin		International Designation	Width		Length	
			Vertical	Horizontal		mm	In.	mm	In.
A (Horiz)	8.5	11.0	0.38	0.25	A4	210	8.27	297	11.69
A (Vert)	11.0	8.5	0.25	0.38	—	—	—	—	—
B	11.0	17.0	0.38	0.62	A3	297	11.69	420	16.54
C	17.0	22.0	0.75	0.50	A2	420	16.54	594	23.39
D	22.0	34.0	0.50	1.00	A1	594	23.39	841	33.11
E	34.0	44.0	1.00	0.50	A0	841	33.11	1189	46.11

1.13 See Figure 1.12. Using a scale of ⅛″=1′–0″, draw the truss shown in the figure. The rise (R) is one-fourth the span of the truss.

Figure 1.12   Problem 1.13   Truss Bridge

1.15 See Figure 1.14. Construct the irregular polygon shown in the figure, using the given dimensions, on an A- or A4-size sheet. Do not dimension.

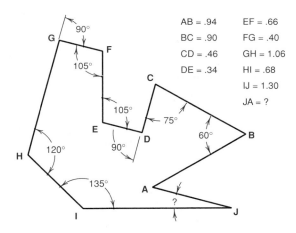

Figure 1.14   Problem 1.15   Irregular Polygon

1.17 See Figure 1.16. Construct the retaining ring shown in the figure. Use an A-size sheet and triple the size of all radii.

Figure 1.16   Problem 1.18   Retaining Ring

1.14 See Figure 1.13. Construct the irregular polygon shown in the figure, using the given dimensions, on an A- or A4-size sheet. Do not dimension.

Figure 1.13   Problem 1.14   Angle Polygon

1.16 See Figure 1.15. Construct the centering plate, using the given dimensions. All of the angles are proportional to angle A. Place the drawing on an A- or A4-size sheet. Do not dimension.

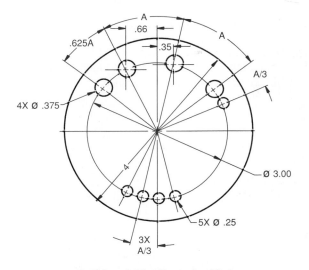

Figure 1.15   Problem 1.16   Centering Plate

1.18 See Figure 1.17. A laser beam directed from source A is reflected at a 45-degree angle from mirror B to mirror C, then onto the horizontal machine surface. Draw the mirrors, machine surface, and light path. Determine angle X for mirror C. [Hint: Angle Y must equal angle Z (angle of incidence equals angle of reflection).] Use a scale of ¼″ equals 1′–0″ and draw on an A-size sheet.

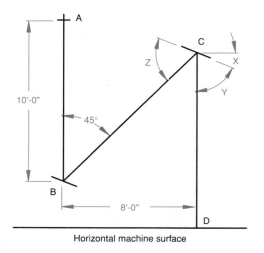

Figure 1.17   Problem 1.17   Reflector

1.19 See Figure 1.18. Construct the pump gasket shown in the figure, using a B-size sheet.

Figure 1.18   Problem 1.19   Pump Gasket

1.20 See Figure 1.19. Construct the chamber clip shown in the figure, using a B-size sheet.

Figure 1.19   Problem 1.20   Chamber Clip

Orthographic Sketch Paper

Sketch Number: _____

Name:_____

Div/Sec: _____

Date:_____

Orthographic Sketch Paper

Sketch Number: _____

Name:_____

Div/Sec: _____

Date:_____

Problem Worksheet

Sketch Number: _____

Name:_____

Div/Sec: _____

Date:_____

Problem Worksheet

Sketch Number: _____

Name:_____

Div/Sec: _____

Date:_____

Sketching and Text

2

OBJECTIVES

After completing this chapter, you will be able to:

1. Define technical sketching.
2. Understand how sketching integrates into the design process.
3. Identify and define two types of sketches.
4. Create a design sketch using pencil or computer.
5. Identify and use sketching tools.
6. Use grid paper to create sketches.
7. Lay out a sketch using proportions.
8. Understand the difference between pictorial and multiview projection.
9. Create a perspective sketch.
10. Create an isometric sketch.
11. Create an oblique sketch.
12. Create a multiview sketch.
13. Identify the types and precedence of lines.
14. Follow good hand-lettering practice.
15. Identify important practices when using CAD for lettering.

2.1 | TECHNICAL SKETCHING

Technical sketching is the process of producing a rough, preliminary drawing representing the main features of a product or structure. Such sketches have traditionally been done freehand; today, CAD systems can also be used. A technical sketch is generally less finished, less structured or restricted, and it takes less time than other types of freehand illustrations. Also, a technical sketch may communicate only selected details of an object, using lines; whole parts of an object may be ignored, or shown with less emphasis, while other features may be shown in great detail.

Technical sketches can take many different forms, depending on the clarity needed and the purpose of the sketch,

Figure 2.1   Technical Sketch

A rough technical sketch can be made to capture a design idea quickly.

both of which depend on the audience for which the sketch is intended. For example, a sketch made quickly to record a fleeting design idea may be very rough. (Figure 2.1) This type of sketch is for personal use and is not meant to be understood by anyone but the individual who produced it. A sketch may also use the format of a more formal, multiview drawing intended to be used by someone who understands technical drawings. (Figure 2.2) However, this type of sketch would not be appropriate for a nontechnical person. Pictorial sketches would be used to further clarify the design idea and to communicate that idea to nontechnical individuals. (Figure 2.3) Shading can be used to further enhance and clarify a technical sketch. (Figure 2.4)

Technical sketches are used extensively in the first (ideation) stage of the design process and are an informal tool used by everyone involved in the design and manufacture of a product. (Figure 2.5) For example, an industrial engineer might make several sketches of a layout for a factory floor.

Many designers find that sketching becomes part of their creative thinking process. Through the process of *ideation,* sketching can be used to explore and solidify de-

Figure 2.2   Multiview Sketch of a Mechanical Part, Used by the Engineer to Communicate Technical Information about the Design to Others

sign ideas that form in the *mind's eye,* ideas that are often graphic in nature. Sketching helps capture these mental images in a permanent form. Each sketch is used as a stepping stone to the next sketch or drawing, where ideas are refined, detail is added, and new ideas are formed.

Figure 2.3 **Pictorial Sketch**
Pictorial sketches are used to communicate technical
information in a form that is easy to visualize.

On a large project, hundreds of sketches are created,
detailing both the successful and the unsuccessful approaches considered for solving the design problem. Since
all but the smallest of design projects are a collaborative
effort, sketches become an important tool for communicating with other members of the design team.

At the early stages of the design process, highly refined, detailed drawings can actually impede the explo-
ration of alternative ideas. What is needed are informal,
nonrestrictive sketches that can communicate both geometric and nongeometric information and can be produced quickly and changed easily. Technical sketching,
being fast and less restrictive, can convey ideas at a level
of detail that communicates the design intent and, at the
same time, allows the viewers to imagine for themselves
how different solutions might further the design. Sketches
as a communications tool encourage collaborative contributions from other members of the design team.

2.1.1 Freehand Sketching Tools

Normally, tools used for sketching should be readily available and usable anywhere: pencil, paper, and eraser. Although variations on these tools are numerous and sophisticated, the goal of technical sketching is simplification. Just a
couple of pencils, an eraser, and a few sheets of paper should
be all that is needed. Many a great design idea was born on
the back of a napkin with a No. 2 wooden pencil! Although
there may be a temptation to use straightedges, such as T-
squares and triangles, a minimum amount of practice should
allow you to draw lines good enough for sketches without
these aids. Mechanically drawn lines can slow you down,
add a level of accuracy not needed in the early stages of a design, and restrict the types of forms explored.

Pencils   The lead used in pencils comes in many different
hardnesses; the harder the lead, the lighter and crisper the
line. For general-purpose sketching, leads in the H and HB

Figure 2.4 **Shaded Sketch**
This rendered sketch is an example of the amount of detail that
can be used when creating sketches. This type of sketch is more
appropriate for technical illustrations than for design
communications. Irwin drawing contest winner Tim Brummett, Purdue
University.

Figure 2.5
Sketching is used throughout the design process to
communicate information.

range will give you acceptable lines. If the lead is much
harder, the lines will be too light and hard to see. In addition, hard lead has a tendency to puncture and tear some of
the lighter weight papers used in sketching. On the other
hand, if the lead is too soft, too much graphite is deposited
on the paper and can be smudged easily. Leads in the middle range allow for a dark, relatively crisp line.

Eraser   Erasing should only be used to correct mistakes
in a line, not to make changes in a design. Such changes

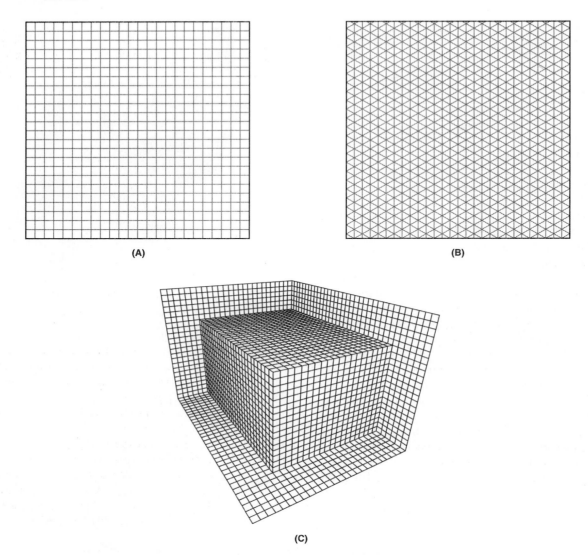

(A)

(B)

(C)

Figure 2.6 Square (A), Isometric (B), and Perspective (C) Grids Used for Sketching
The grid lines are used as an aid in proportioning the drawing and sketching straight lines freehand.

should be recorded on a separate sketch, and the original sketch should be preserved. Still, most people find that a small amount of erasing is helpful. Usually, the eraser on the end of the pencil is sufficient. However, if you are going to do a lot of sketching, you may need a separate eraser, and one of any size or shape will do. You might consider a gum eraser, since they leave less residue when used.

Paper   There is a wide range of paper choices for sketching (including a napkin you could draw on during lunch). The most accessible and easiest to use is notebook size (8-½″ × 11″) paper. Because of the difficulty of drawing long lines freehand, paper much larger than that is normally not useful for a single sketch. On the other hand, larger paper is useful for drawing multiple sketches that should be visually grouped together.

Plain bond paper with no lines offers the highest degree of flexibility; lined paper tends to lock you in visually to drawing along the lines. However, when you want the guidance of existing lines on the paper, it is most useful to have the lines running along both dimensions, forming a grid. Two of the most common **grid papers** used in sketching are **square grid** (Figure 2.6A) and **isometric**

grid (Figure 2.6B) for use in certain types of pictorial sketch. Common grid densities run from 4 to 10 lines per inch. A less common type of grid paper is perspective, which is used to create another type of pictorial sketch. (Figure 2.6C)

Often, it would be useful to have grid lines for the sketch, but not for the final drawing. One way this can be achieved is to sketch on thin, plain, semitransparent **tracing paper** laid over the grid paper and taped down so that the grid lines show through. When the sketch is done, it is untaped from the grid paper and viewed without the grid lines behind it. This technique is also a money saver because grid paper is more expensive than tracing paper (often called trash paper), which can be bought in bulk on rolls. The other advantage to tracing paper is that it can be laid over other sketches, photos, or finished technical drawings. A light table can be used to improve the tracing process. Tracing is a fast, accurate method for refining a design idea in progress or for using an existing design as the starting point for a new one.

2.2 | SKETCHING TECHNIQUE

It takes practice and patience to produce sketches that are both legible and quickly made. The following sections describe common techniques used to produce good sketches quickly. The discussions cover the tools and the techniques for creating straight lines, curves (such as circles and arcs), and proportioned views. With patience and practice, it is possible for you to become good at making quick, clear sketches, regardless of your experience and natural drawing ability.

2.2.1 Seeing, Imaging, Representing

There are certain fundamental skills that must be learned in order for sketching to be used as a tool for design. Over a period of time and with practice you will be able to acquire the skills and knowledge necessary to create design sketches. Sketching is based on seeing (perception) and visual thinking through a process of seeing, imaging, and representing (Figure 2.7). Seeing is our primary sensory channel because so much information can be gathered through our eyes. It is our best developed sense and one we take for granted every day as we easily move through our environment. Seeing empowers us to sketch.

Imaging is the process used by the mind that takes the visual data received by our eyes to form some structure

Figure 2.7 The Sketching Process
Sketching is based on the interactive process of seeing, imaging, and representing.

and meaning. The mind's eye initially creates the images whether real or imagined, and these are the images used to create sketches. **Representing** is the process of creating sketches of what our minds see.

Seeing and imaging is a pattern-seeking process where the mind's eye actively seeks those features that fit within our interests, knowledge, and experiences. Figures 2.8 and 2.9 are examples of sketches that can be interpreted in more than one way. It is also possible to make sketches of objects that cannot exist in the real world. M.C. Escher was a genius at creating sketches and drawings of objects or environments that could not exist in reality (Figure 2.10).

Practice Exercise 2.1

Our perception is not limited to what we can see. Images often appear spontaneously in response to a memory recall. In this exercise read the words and see if visual images are created in your mind's eye.

1. Your bedroom where you grew up as a child, or the street you lived on.

2. A close relative, a famous actor, or a close friend from high school.

3. A basketball sitting at center court on your high school gym floor. Try sketching the basketball on the floor.

Your response to these written prompts is an example of your visual memory. You are thinking visually, which is a very powerful way of thinking when designing.

Figure 2.8   Pattern-seeking Process of the Human Mind
In the illusion on the left, created by psychologist E. G. Boring in 1930, you can see either the head of an older woman or the profile of a younger woman. The illustration on the right can be viewed as either a vase or two profiles of the same person looking at each other.

2.2.2   Contour Sketching

The most fundamental element to creating sketches is the line or the outline of an object. The lines or outlines of an object are used to represent the edges and contours of objects we see in the world. If we sketch the boundaries, an object slowly takes shape and we begin to recognize it as a familiar object. This technique of sketching the outline of an object is called **contour sketching** and is an important technique used by novice sketchers to gain confidence in their sketching ability. Contours on objects can take the form of edges of an object, lines that separate contrasting light or color, changes in the surface of an object, and overlapping parts. The primary reason for contour sketching is to develop your visual acuity and sensitivity to important object features which are necessary to create accurate sketched representations.

When first learning how to use contour sketching, begin by slowly tracing the outline of an object with your eyes while slowly sketching what you see. At first the sketch may seem crude and out of proportion, but with practice your sketches will be quite good. Figure 2.11 shows an example of a contour sketch created by carefully looking at the outline of the object and sketching what you see without looking at the paper. Figure 2.12 is a sketch created by care-

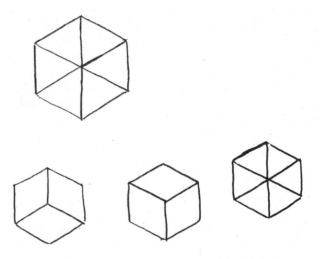

Figure 2.9   Different Interpretations of the Same Object
In this example a cubelike illustration can be interpreted as a hexagonlike figure or as a cube viewed from below, above, or with transparent sides.

fully looking at the outline of the object and looking at the paper as you sketch. Both techniques are useful when learning how to observe and create sketches of what you see.

Figure 2.10 Optical Illusion
Follow the path of the water in this illustration to see the optical illusion created by M. C. Escher. You can also see Escher's early design sketches of the waterfall.   (M.C. Escher's "Waterfall" © 2000 Cordon Art B.V. -Baarn-Holland. All rights reserved.)

Making a Contour Sketch

In this exercise, you are to create a sketch of the stapler shown in Figure 2.11 using the contour sketching technique.

Step 1.   Using a plain piece of white paper and a soft lead pencil, place your drawing hand with the pencil near the center of the paper.

Step 2.   Orient the paper in a comfortable position for sketching.

Step 3.   Comfortably and in a relaxed manner, very slowly begin to trace the outline of the object with your eyes.

Step 4.   Slowly move you pencil across the paper as your eye scans the outline of the object. Do not erase or sketch over lines and do not look at your sketch. Sketch very slowly and deliberately.

Step 5.   Continue to draw each edge as you view it at a slow and deliberate pace.

Step 6.   Look at your sketch after you have completed viewing the contours of the object.

Figure 2.11   Contour Sketch

A contour sketch is created by carefully observing the outline of an object while sketching. This technique is used to improve your sketching ability. In this example, the contour sketch was created without looking at the paper.

Figure 2.12   Modified Contour Sketch

This contour sketch was created by looking at the object, then looking at the paper as the sketch was produced.

Making a Modified Contour Sketch

In this exercise, you are to create a contour sketch but you will be able to look at your sketch as you are working (Figure 2.12).

Step 1.   Using a plain piece of white paper and a soft lead pencil, place your drawing hand with the pencil near the center of the paper.

Step 2.   Orient the paper in a comfortable position for sketching.

Step 3.   Comfortably and in a relaxed manner, very slowly begin to trace the outline of the object with your eyes.

Step 4.   Slowly move you pencil across the paper as your eyes scans the outline of the object. Do not erase or sketch over lines. Sketch very slowly and deliberately.

Step 5.   Occasionally look at your sketch to match it with the object being drawn.

Step 6.   Continue to draw each edge and interior edges as you view it at a slow and deliberate pace.

2.2.3   Negative Space Sketching

Another useful technique novice sketchers can try to improve their sketching technique is called **negative space sketching**. In this technique you concentrate on the

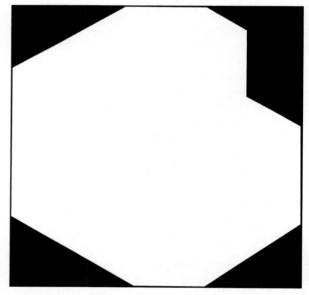

Figure 2.13   Negative Space Sketching

Negative space sketching produces a sketch having only the spaces between the object and not the object itself.

spaces between the objects and not on the object itself. In other words, you concentrate on the geometry of the object, such as lines, curves, angles, tangencies, and not the names of the objects, such as handle, hole, base, cube. An example of a negative space sketch is shown in Figure 2.13. Notice the object itself is not shaded and lacks details, but the space surrounding the object is shaded.

Making a Negative Space Sketch

For this exercise, you are to create a negative space sketch of the object shown in Figure 2.13.

Step 1. Use a plain sheet of white paper and begin by sketching the box surrounding the object.

Step 2. Sketch over the top of the negative spaces in the figure to reinforce that you are going to be sketching the negative spaces and not the object itself.

Step 3. Focus on one of the outlined negative spaces just created in step 2 until you can visualize the negative space.

Step 4. Now begin sketching the negative space form on your sheet of paper. Concentrate on drawing lines and curves by determining the angles, lengths, tangencies, and other geometric characteristics.

Step 5. Repeat steps 3 and 4 until all the negative space has been created.

2.2.4 Upside-Down Sketching

Upside-down sketching is another method that can be used to improve your sketching ability. In this technique you take a photograph of a recognizable object, such as a chair, and turn it upside-down before sketching it. By turning it upside-down you can concentrate on the shape and form of the object, allowing you to create a better sketch. Figure 2.14 is a photograph of a table that is upside-down. Carefully sketch the outline of the object by concentrating on the geometry or form and not the names of the part, such as legs or feet. By doing so you will be able to create a more accurate sketch of the object.

2.2.5 Straight Lines

All sketches are made up of a series of lines. Lines created for sketches differ from mechanically produced lines in that they are not constrained or guided by instruments, such as a T-square, template, or compass. Instead, the lines are guided strictly by the eye and hand. Such lines have a different aesthetic quality than mechanical lines. (Figure 2.15) At a micro level, sketched straight lines are uneven; at a macro level, they should appear to follow a straight path without any interruptions. (Figure 2.16)

One of the easiest guides to use for sketched lines is grid paper. Lines drawn right on the grid are the easiest to produce, and even those lines that are offset but parallel to a grid line are fairly easy to produce. The idea is to keep your sketched line a uniform (but not necessarily equal) distance between two existing grid lines.

Curved lines, straight lines not parallel to a grid line, and lines drawn without the aid of a grid are more difficult. In all

Figure 2.14   Upside-Down Sketching
Sketch the outline of the object by concentrating on the geometric forms.
(Courtesy of Lunar Design Incorporated.)

Mechanical

Sketched

Figure 2.15   A Comparison of Mechanically Drawn and Sketched Lines

Yes

No!

No!

No!

Figure 2.16   Examples of Good and Bad Straight Line Technique
Sketched lines should be straight and dark and should have a consistent thickness.

of these cases, the lines are drawn as interpolations between two or more points. The points are typically marked on an engineering drawing as two intersecting lines, one horizontal and one vertical, and each approximately $\frac{3}{16}''$ long. Your eye

Figure 2.17 Sketching Lines
The sequential drawing of a straight line is done by first drawing a very light line, using short strokes. The light line is then drawn over and darkened.

should take a "global" view of all the points to be connected and should guide your hand as it goes from point to point.

Quite often, the sketched line is built up from a sequence of two or three passes with the pencil. (Figure 2.17) The first pass is drawn light, using a hard lead, such as a 4H, sharpened to a point, and may not be as straight as your final line will be; however, it should provide a path on top of which the final, even, darker line is drawn. For particularly long lines, the initial line may be drawn in segments, coming from the two endpoints and meeting in the middle; however, the final line should be drawn in one single pass to avoid choppiness. If necessary, another pass can be used to darken or thicken the line.

Long lines are difficult to control, even for someone with a lot of experience. If you cannot choose a drawing scale that reduces the size of the sketch, use grid paper as a guide, drawing either directly on the grid paper or on tracing paper placed on top of the grid paper. If the line is parallel and relatively close to the edge of the paper, you can rest a finger or a portion of your palm along the edge of the paper to sta-

bilize your drawing hand. (Figure 2.18) If necessary, you can use a ruler or a scrap of paper to mark a series of points on the sketch, but this will slow you down a bit.

Another technique that helps when drawing lines of any length is changing the orientation of the paper. Sketching paper should not be fixed to your drawing surface. Instead, you should be able to rotate the paper freely, orienting it in the direction that is most comfortable. Practice will determine which orientation is best for you. Many people find that drawing the lines by moving away from or toward the body, rather than from left to right, produces the quickest, straightest lines; others find it most comfortable if the paper is angled slightly away from the body. Again, the longer the line, the more important it is that the paper be positioned comfortably for you.

The following summarizes the techniques used to sketch straight lines:

- Orient the paper to a comfortable position. Do not fix the paper to the surface.
- Mark the endpoints of the lines to be sketched.
- Determine the most comfortable method of creating lines, such as drawing from left to right, or drawing either away from or toward your body.
- Relax your hand and the rest of your body.
- Use the edge of the paper as a guide for making straight lines.
- Draw long lines by sketching a series of connected short lines.
- If necessary, draw on grid paper or on tracing paper that is overlaid on grid paper.

Figure 2.18 Sketching Long Lines
Very long lines can sometimes be more difficult to sketch. One technique is to use the edge of the paper as a guide for your hand. Another technique is to mark equal distances from the edge of the paper using a marked scrap of paper as a guide. The marks are then used as a guide to produce the line.

Sketching Straight Lines

In this exercise, you are to create a series of 5″ long parallel lines equally spaced at 0.5″. Refer to Figures 2.16 and 2.17.

Step 1. Lightly mark the endpoints of the lines to be sketched on 8-½″ × 11″ paper.

Step 2. Orient the paper in a comfortable position for sketching.

Step 3. Comfortably and in a relaxed manner, position your hand so that the pencil is close to one of the marked endpoints of the first line to be sketched. Sketch the top line first, to avoid smearing newly sketched lines with your hand.

Step 4. Quickly scan the two endpoints of the first line to determine the general direction in which you will be sketching.

Step 5. Lightly sketch a short line, approximately 1″ long, by moving your hand and the pencil in the general direction of the other end of the line.

Step 6. Repeat steps 4 and 5 until the other end of the line is reached.

Step 7. Return to the starting point of the line and overdraw the line segments with a slightly longer, heavier stroke, to produce a thick, dark, more continuous straight line.

Step 8. Repeat steps 3 through 7 to sketch the remaining straight lines.

2.2.6 Curved Lines

Curved lines need multiple guide points. The most common curve is a circle or circular arc. Although very small circles and arcs can be drawn in one or two strokes and with no guide points, larger circles need some preliminary points. The minimum number of points for a circle is four, marked on the perimeter at equal 90-degree intervals. For an arc, use at least one guide point for every 90 degrees and one at each end.

There are a number of ways to lay out the guide points for circular curves quickly. One way is to draw a square box whose sides are equal to the diameter of the circle. (Figure 2.19A) The midpoints on each side of the square mark the points where the circle will touch the square. These points are called points of tangency. More guide points can be added by drawing the two diagonals across the square. The center of the circle being sketched is the point where the diagonals cross. (Figure 2.19B) Mark the guide points on each diagonal approximately two-thirds the distance from the center of the circle to the corner of the square. This distance is the approximate radius of the circle. (Figure 2.19C)

As with longer straight lines, large arcs and circles are harder to draw and may need guide points marked at more frequent intervals. To do this, it is handy to use a scrap of paper with the radius marked on it. (Figure 2.19D)

Circular arcs are drawn the same way as circles, adjusting the number of points to suit the degree of curvature (i.e., the length) of the arc. Noncircular arcs, however, can be more difficult. Since these lines are only to be sketched, calculating the points that the curve should pass through is too involved and is not recommended. Simply use the eye to estimate guide points and then gradually draw a curve to pass through those points. (Ellipses and curves in multiview drawings are two special cases treated later in this chapter.)

As with straight lines, positioning the paper and using a relaxed grip are important for helping you create good curves. Unlike straight lines, curves are usually best drawn in a series of arcs of not more than 90 degrees. After each arc is drawn, rotate the paper for the next segment of arc. With practice you may be able to eliminate rotating the paper for smaller arcs, but will probably still have to do so for larger ones.

Sketching a Circle or Arc

The following steps demonstrate how to sketch a circle or arc. Refer to Figures 2.19 as a guide.

Step 1. Orient the paper in a comfortable position and relax your grip on the pencil. Lightly mark the corners of a square with sides equal in length to the diameter of the circle or arc to be sketched.

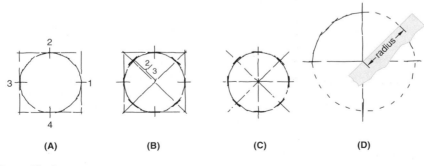

(A) (B) (C) (D)

Figure 2.19 Sketching a Circle
Sketching a circle is easier using one of the techniques shown. For small circles, use a square or multiple center lines to guide the construction process. For large circles, use a scrap of paper with the radius marked on it as a guide.

Step 2. Lightly sketch the square, using short strokes to create the straight lines.

Step 3. Mark the midpoints of the four sides of the square. This gives you four marks on the perimeter of the circle.

Step 4. Sketch diagonals across the corners of the square. Where the diagonals cross is the center of the circle.

Step 5. Mark the diagonals at two-thirds the distance from the center of the circle to the corner of the square. This gives you four more marks on the circle's perimeter.

Step 6. Sketch the circle by creating eight short arcs, each between two adjacent marks on the perimeter. Start at any mark and sketch an arc to the next mark (on either side of the first one, whichever is most comfortable for you).

Step 7. Rotate the paper and sketch the next arc from the last mark you touched to the next mark on the perimeter. Repeat this step until all eight arc segments have been sketched. For smoother sketches, rotate the paper in the opposite direction from the one you used to draw the arc.

Step 8. Overdraw the arcs with a thick, black, more continuous line to complete the sketched circle.

2.3 | PROPORTIONS AND CONSTRUCTION LINES

Frequently, in the sketch of an object, the relative proportions of its primary dimensions—width, height, and depth—are more important than their actual physical sizes. A **proportion** is the ratio between any two dimensions of an object. These proportions are represented in the sketch by a series of preliminary lines, which are drawn light and fast, and which may or may not represent the locations of the final lines in the sketch. Their purpose is to form a backbone, a structure inside which the final linework can be drawn.

The first step in a sketch involves drawing the construction lines, which guide a sketch's overall shape and proportion. **Construction lines** are very light, thin lines used to roughly lay out some of the details of sketches or drawings. Do not try to draw the construction lines to exact lengths since lengths are marked later, either by intersecting lines or short tick marks.

Construction lines have two primary features: the lines themselves and the intersections created where two lines cross. For example, the construction lines become the paths for the final straight lines. Points marked by the intersections of construction lines guide the drawing of circles. Usually, both of these features are used in creating sketches. Since all the dimensions of a sketch are estimated, groups of construction lines forming boxes and other shapes are an important tool for preserving the shape and proportion of the object and its features as the sketch is developed.

Grid paper can be used as a guide in creating construction lines but should not be thought of as a substitute, since the grid does not directly represent the proportions of the object, and there are many more grid lines than there are features on the object. The goal is to draw construction lines on top of the grid to reveal the form of the object. With experience, you may be able to make do with fewer construction lines, but while you are learning how to create properly proportioned sketches, you should use more, rather than fewer, construction lines to guide you.

Each feature has a proportion that can be represented by a series of construction lines. The following steps describe how to proportion a drawing by breaking it down into its component features.

Creating a Proportioned Sketch

Step 1. Refer to Figure 2.20. Gage the proportion of the overall size of the object. For the first sketch, use two overall dimensions of the object: width and height. Lightly sketch a box that represents the ratio of these two dimensions. (Figure 2.20, step 1) This box is called a **bounding box** because it represents the outer dimensional limits of the feature being drawn. If the object is rectangular in shape, the final linework will follow the perimeter of the bounding box. In most cases, however, the final linework will only touch on a portion of the box's edges.

Step 2. Inside the first bounding box, draw other boxes to represent the larger features of the object, and within those boxes draw still others to represent the smaller features of the object. Often, a construction line can be used for more than one box. The final boxes each show the proportions of one feature of the object.

Step 3. Continue to draw bounding boxes until the smallest features of the object have been represented. As you gain experience, you may find that some of these smaller features need not be boxed; instead, their final lines can be sketched directly.

Step 4. When all of the features of the object have been boxed, begin sketching the final linework, which is done significantly darker than the construction lines.

The goal is, if you hold the drawing at arm's length, the construction lines are hard to see, and the final linework is clear and sharp. If there is not enough contrast between the construction lines and the final linework, then the construction lines become a distraction. Make the final lines darker, or the construction lines lighter, or both; however, do not erase your construction lines.

One of the most difficult sketching techniques to learn is making a sketch look well proportioned. For example, Figure 2.21 shows a well proportioned and a poorly pro-

Figure 2.20   **Creating a Proportioned Sketch**
To create a well-proportioned sketch, use multiple steps to create lightly sketched rectangles and squares that are then used as guides for the final sketch.

portioned sketch of a computer monitor. Proportioning skills will improve with practice. A good rule of thumb is, if the drawing does not look or feel right, it probably is not. In the poorly proportioned monitor in Figure 2.21, the ratio of the height to the width is incorrect.

Sketching Objects

Step 1. Collect magazine photographs or clippings that show 2-D images or patterns. These can range from pictures of faces, to company logos, to fronts of buildings, etc. Stick with images that look *flat;* that is, images that don't show a depth dimension.

Step 2. Lay tracing paper over the image and tape the paper down.

Step 3. Lightly sketch an overall bounding box of the object. Look at the image contained in the bounding box. Mentally identify as many features on the object as you can. The features may be small and self-contained or a collection of several smaller features.

Step 4. Refine the drawing by sketching a series of progressively smaller bounding boxes. Start with the larger features and work downward.

Figure 2.21   **Good and Poor Proportions**
One well and one poorly proportioned sketch of a computer monitor. The poorly proportioned monitor looks too wide.

Step 5. If desired, you can then darken some of the lines representing the image, to highlight the most important lines of a feature. What are the most important lines of a feature? Experiment with different lines to see which are more critical than others in representing the form of the image.

Hint: Buy a roll of tracing paper from your local blueprint or art supply store. It's cheaper than individual sheets, and you won't run out as often.

2.4 | INTRODUCTION TO PROJECTIONS

Both ideation and document sketches can represent the objects being designed in a number of different ways. We live in a three-dimensional (3-D) world, and representing that world for artistic or technical purposes is largely done on two-dimensional (2-D) media. Although a sheet of paper is technically three-dimensional, the thickness of the paper (the third dimension) is useless to us. It should be noted that the computer screen is a form of two-dimensional media, and images projected on it are governed by the same limitations as projections on paper. Modern techniques, such as holograms, stereograms, and virtual reality devices, are attempts to communicate three-dimensional ideas as three-dimensional forms. However, drawings are still the primary tool used for representing 3-D objects.

Most projection methods were developed to address the problem of trying to represent 3-D images on 2-D media. (Figure 2.22) Projection theory and methods have taken hundreds of years to evolve, and engineering and technical graphics is heavily dependent on projection theory.

The most common types of projection used in sketching are *multiview, isometric* (one type of axonometric), *oblique,* and *perspective,* as shown in Figure 2.23. These four types of projection can be placed in two major categories: multiview sketches and pictorial sketches. **Multiview sketches**

2-D Paper

3-D Part

Figure 2.22 3-D Object on 2-D Medium
For centuries, graphicians have struggled with representing 3-D objects on 2-D paper. Various projection techniques have evolved to solve this problem.

(A) Multiview

(B) Axonometric

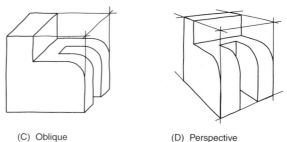

(C) Oblique

(D) Perspective

Figure 2.23 Classification of Sketches
Various projection techniques are used to create four basic types of sketches: multiview, axonometric, oblique, and perspective. The sketches shown in B, C, and D are called pictorial because they represent the object as a 3-D form. The multiview sketch uses multiple flat views of the 3-D object to accurately represent its form on 2-D paper.

present the object in a series of projections, each one showing only two of the object's three dimensions. The other three types of projection, grouped as **pictorial sketches,** present the object in a single, pictorial view, with all three dimensions represented. There are always trade-offs when using any type of projection; some are more realistic, some are easier to draw, and some are easier to interpret by non-technical people.

Axonometric projection is a parallel projection technique used to create a pictorial drawing of an object by rotating the object on an axis relative to a *projection,* or *picture plane.* In multiview, axonometric, and oblique projection, the observer is theoretically infinitely far away from the projection plane. In addition, for multiviews and axonometric projections the lines of sight are perpendicular to the plane of projection; therefore, both are considered *orthographic projections.* The difference s between multiview drawing and an axonometric drawing are that, in a multiview, only two dimensions of an object are visible on each view and more than one view is required to define the object, whereas in an axonometric drawing, the object is rotated about an axis to display all three dimensions, and only one view is required.

Axonometric drawings are classified by the angles between the lines comprising the axonometric axes. The **axonometric axes** are axes that meet to form the corner of the object that is nearest to the observer.

When all three angles are unequal, the drawing is classified as a **trimetric projection**. When two of the three angles are equal, the drawing is classified as a **dimetric projection**. When all three angles are equal, the drawing is classified as an **isometric (equal measure) projection**.

Mechanically drawn pictorials can often be as hard to draw as multiviews. Various 2-D CAD-based tools have eased the process of creating pictorials. Probably the easiest way of creating such views is to use a 3-D CAD package to create a model. This model can easily represent pictorial views and can also generate views for a multiview drawing.

Another way of classifying projections relates to whether they use **parallel projection** or **perspective projection.** Multiview, isometric, and oblique multiview projections use parallel projection, which preserves the true relationships of an object's features and edges. This type of projection is the basis of most engineering and technical graphics. Perspective projection distorts the object so that it more closely matches how you perceive it visually.

Since it is much easier to lay out a sketch in parallel rather than in perspective projection, you will probably

Figure 2.24   Perspective Sketch

For perspective projection, the width and depth dimensions converge on vanishing points.

find yourself doing a majority of your sketching using parallel projection, even though it is less realistic. Only when the object spans a large distance—such as a house or bridge—will it be useful to represent the distortion your eyes perceive as the object recedes from view.

2.4.1 Isometric Pictorials

An **isometric pictorial** sketch is a type of parallel projection that represents all three dimensions in a single image. Although there are a number of ways of orienting an object to represent all three dimensions, isometric pictorials have a standard orientation that makes them particularly easy to sketch. Start by looking at the two-point perspective in Figure 2.24. Then, instead of having the width and depth construction lines converge on vanishing points, have them project parallel to each other at a 30-degree angle above the baseline. (Figure 2.25)

Many CAD systems will automatically produce an isometric view of a 3-D model when the viewing angle is specified. Some CAD systems have predefined views, such as isometric, which are automatically created after selection.

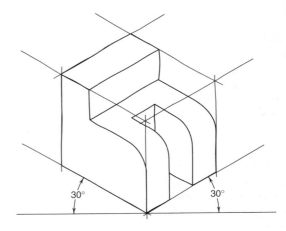

Figure 2.25   Isometric Sketch

For this isometric sketch, the width and depth dimensions are sketched 30 degrees above the horizontal.

struction line through the bottom of the vertical line. Sketch a line from the base of the vertical line to the right, at an approximate angle of 30 degrees above the horizontal construction line. Sketch a line from the base of the vertical line to the left, at an approximate angle of 30 degrees above the horizontal construction line.

The corner of the axis is labeled point 1; the end of the width line is labeled point 2; the end of the depth line is labeled point 4; and the top of the height line is labeled point 3. The lengths of these lines are not important, since they will be treated as construction lines, but they should be more than long enough to represent the overall dimensions of the object. Estimate the overall width, height, and depth of the object using the estimating techniques described earlier in this chapter. Use these dimensions to sketch a block that would completely enclose the object.

Blocking in the object

Step 3. Sketch in the front face of the object by sketching a line parallel to and equal in length to the width

Making an Isometric Sketch

Make an isometric sketch of the object shown in Figure 2.26.

Sketching the isometric axis

Step 1. Isometric sketches begin with defining an **isometric axis,** which is made of three lines, one vertical and two drawn at 30 degrees from the horizontal. These three lines of the isometric axis represent the three primary dimensions of the object: width, height, and depth. Although they are sketched at an angle of only 60 degrees to each other, they represent mutually perpendicular lines in 3-D space.

Step 2. Begin the sketch by extending the isometric axes shown in Step 1, Figure 2.26. Sketch a horizontal con-

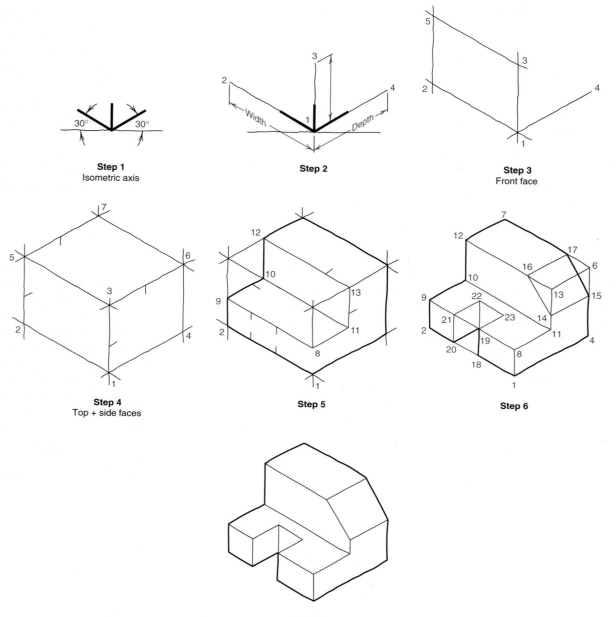

Step 1
Isometric axis

Step 2

Step 3
Front face

Step 4
Top + side faces

Step 5

Step 6

Step 7

Figure 2.26 The Basic Steps Used to Create an Isometric Sketch of an Object

dimension, passing the new line through point 3. Sketch a line parallel to and equal in length to the vertical line (1–3), through points 5–2. The front face of the object is complete.

Step 4. From point 3, block in the top face of the object by sketching a line parallel to and equal in length to line 1–4. This line is labeled 3–6. Sketch a line parallel to and equal in length to line 3–5, from point 6. This line is labeled 6–7.

Sketch a line from point 5 to point 7. This line should be parallel to and equal in length to line 3–6. Block in the right side face by sketching a line from point 6 to point 4, which is parallel to line 1–3. The bounding box of the object, sketched as construction lines, is now finished. The box serves the same purpose as the one drawn in Figure 2.12, but it represents all three dimensions of the object instead of just two.

Adding details to the isometric block

Step 5. Begin by estimating the dimensions to cut out the upper front corner of the block, and mark these points as shown in Step 4. Sketch the height along the front face by creating a line parallel to line 1–2; label it 8–9. Sketch 30-degree lines from points 8 and 9 and label these lines 9–10 and 8–11. Now sketch a line from point 10 to point 11. Sketch vertical lines from points 10 and 11 and label the new lines 10–12 and 11–13. Sketch a line from point 12 to point 13, to complete the front cutout of the block.

With a simple sketch, you can often lay out all of your construction lines before having to darken in your final linework. With more complicated sketches, the sheer number of construction lines can often cause confusion as to which line belongs to which feature. The confusion can be worse in an isometric sketch, where the lines represent three dimensions rather than two. Therefore, after the marks are made for the last two features in step 5, you can begin darkening in some of the lines representing the final form.

Step 6. Estimate the distances to create the angled surface of the block, and mark these points, as shown in step 5. From the marked point on line 11–13, sketch a 30-degree line to the rear of the block on line 4–6. Label this new line 14–15. From the marked point on line 12–13, sketch a 30-degree line to the rear of the block on line 6–7. Label this new line 16–17. Sketch a line from point 14 to point 16 and from point 15 to point 17, to complete the sketching of the angled surface. Lines 14–16 and 15–17 are referred to as **nonisometric lines** because they are not parallel to the isometric axis.

Estimate the distances for the notch taken out of the front of the block, and mark these points, as shown in step 5. Draw vertical lines from the marked points on line 1–2 and line 8–9. Label these lines 18–19 and 20–21, as shown in step 6. Sketch 30-degree lines from points 19, 20, and 21 to the estimated depth of the notch. Along the top surface of

the notch, connect the endpoints of the 30-degree lines, and label this new line 22–23. The 30-degree line extending back from point 20 is stopped when it intersects line 18–19, as shown in step 6. To complete the back portion of the notch, drop a vertical line from point 22, as shown in step 6. Stop this new line at the intersection point of line 19–23. The rough isometric sketch of the block is complete.

Note that we have not mentioned hidden features representing details behind the visible surfaces. The drawing convention for isometric sketches calls for disregarding hidden features unless they are absolutely necessary to describe the object.

Step 7. Darken all visible lines to complete the isometric sketch. Since the construction lines are drawn light, there is no need to lighten them in the completed sketch.

2.4.2 Isometric Ellipses

Isometric ellipses are a special type of ellipse used to represent holes and ends of cylinders in isometric drawings. In an isometric drawing, the object is viewed at an angle, which makes circles appear as ellipses. When sketching an isometric ellipse, it is very important to place the major and minor axes in the proper positions. Figure 2.27 is an isometric cube with ellipses drawn on the three visible surfaces: top, profile, and front. Remember Figure 2.27A, because those are the three positions of isometric ellipses found on most isometric sketches and drawings. The following are the key features of the isometric ellipse on each plane:

- The major and minor axes are always perpendicular to each other.
- On the top plane, the major axis is horizontal, and the minor axis is vertical.

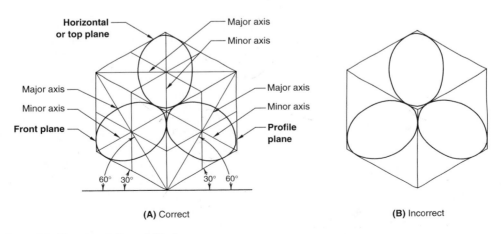

(A) Correct   **(B) Incorrect**

Figure 2.27   Isometric Representation of Circles
Circular features appear as ellipses in isometric sketches. The orientation of the ellipse is set according to the face on which the circle lies. The correct orientation is shown in (A) and examples of incorrect orientations are shown in (B).

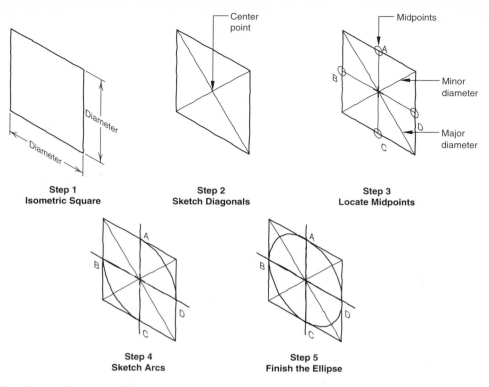

Step 1
Isometric Square

Step 2
Sketch Diagonals

Step 3
Locate Midpoints

Step 4
Sketch Arcs

Step 5
Finish the Ellipse

Figure 2.28   Sketching an Isometric Ellipse

The steps used to create a sketch of an isometric ellipse begin with constructing an isometric box whose sides are equal to the diameter of the circle. The center of the box and the midpoints of the sides are found, and arcs are then drawn to create the ellipse.

- On the front and profile planes, the major axes are measured 60 degrees to the horizontal.
- The major axis is always perpendicular to the axis running through the center of the hole or cylinder.

Sketching an Isometric Ellipse

Figure 2.28 shows the steps for creating an isometric ellipse. Notice that the steps are almost identical to those for sketching a circle as explained earlier in this chapter. The difference is in the orientation and proportion of the primary axes.

Step 1. This isometric ellipse will be drawn on the front plane. Begin by sketching an isometric square whose sides are equal to the diameter of the circle.

Step 2. Add construction lines across the diagonals of the square. The long diagonal is the **major axis,** and the short diagonal is the **minor axis** of the ellipse. The two diagonals intersect at the center of the square, which is also the center of the isometric ellipse.

Step 3. Sketch construction lines from the midpoints of the sides of the square through the center point. These lines represent the center lines for the isometric ellipse. The midpoints of the sides of the isometric square will be tangent points for the ellipse and are labeled points A, B, C, and D.

Step 4. Sketch short, elliptical arcs between points B and C and points D and A.

Step 5. Finish the sketch by drawing the elliptical arcs between points C and D and points A and B, completing the ellipse.

Sketching an Isometric Cylinder

Figure 2.29 shows the steps for creating an isometric view of a cylinder.

Step 1. Sketch the isometric axis. To sketch the bounding box for the cylinder, begin on one 30-degree axis line and sketch an isometric square with sides equal to the diameter of the cylinder. This square will become the end of the cylinder. Next, mark the length of the cylinder on the other 30-degree axis line, and sketch the profile and top rectangles of the bounding box. For the profile rectangle, the length represents the length of the cylinder, and the height represents the diameter of the cylinder. For the top rectangle, again the length represents the length of the cylinder, but the width represents the diameter of the cylinder. Note that only three long edges of the bounding box are drawn (the hidden one is not), and only two lines for the back end of the bounding box are drawn (the two hidden ones are not).

Step 2. Draw diagonals and center lines on the isometric square, and sketch in the ellipse, to create the end of the cylinder, as described in "Sketching an Isometric Ellipse."

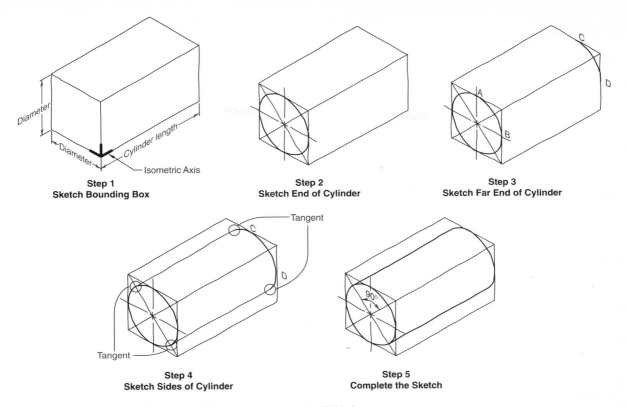

Figure 2.29 **Steps Used to Construct a Sketch of an Isometric Cylinder**

Step 3. Where the center lines intersect with the top and front sides of the isometric square, mark points A and B. Sketch construction lines from points A and B to the back end of the bounding box and mark points C and D. Sketch an arc between points C and D.

Step 4. On the isometric square, locate the two points where the long diagonal intersects with the ellipse. From those two points, sketch two 30-degree lines to the back of the bounding box. (These 30-degree lines are tangent to the ellipse on the front of the cylinder.) Then, sketch short elliptical arcs from points C and D to the tangent lines, as shown in the figure. The cylinder should now be visible in the bounding box.

Step 5. Darken all visible lines to complete the cylinder. Note that the major axis of the ellipse is perpendicular to the axis running through the center of the cylinder, and the minor axis is coincident to it.

Sketching Semi-Ellipses

Figure 2.30 shows how to sketch a semi-ellipse.

Step 1. This isometric ellipse will be drawn on the profile plane. Begin by sketching an isometric square whose sides are equal to the diameter of the arc. Add construction lines across the diagonals of the square. The two diagonals intersect at the center of the square, which is also

the center of the isometric ellipse. Sketch construction lines from the midpoints of the sides of the square through the center point. These lines represent the center lines for the isometric ellipse.

Step 2. The midpoints of the sides of the isometric square will be tangent points for the ellipse and are labeled points A, B, and C. The long diagonal is the major axis, and the short diagonal is the minor axis. Sketch short, elliptical arcs between points B and C and points B and A, which creates the elliptical arc on the near side of the object. The back part of the semi-ellipse can be sketched by constructing 30-degree parallel lines that are equal in length to the depth of the part, from points A, B, and C. This locates points A', B', and C' on the back side of the object. Add the back ellipse by sketching an arc between points B' and C' and points B' and A'.

Step 3. Finish by darkening the final lines and lightening the construction lines.

2.4.3 Isometric Grid Paper

The use of isometric grid paper can improve your technique and decrease the time necessary to create an isometric sketch. Isometric grid paper is made of vertical and 30-degree grid lines, as shown in Figure 2.6B. There are two primary advantages to using isometric grid paper. First,

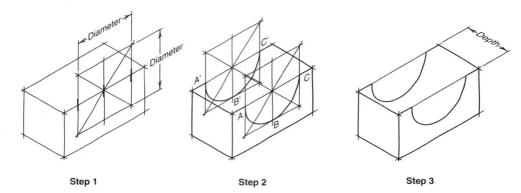

Figure 2.30   Steps Used to Construct a Sketch of a Semi-Ellipse

there is the advantage obtained by using any kind of grid paper. Proportions of the object's features can be translated into certain numbers of blocks on the grid. This can be especially useful when transferring the dimensions of a feature from one end of the object to the other. Unlike square grid paper, each intersection on an isometric grid has three lines passing through it, one for each of the primary axis lines. This can create some confusion when counting out grid blocks for proportions. Just remember which axis line you are using and count every intersection as a grid block.

The second advantage of the isometric grid is the assistance it provides in drawing along the primary axis lines. Although estimating a vertical line is not difficult, estimating a 30-degree line and keeping it consistent throughout the sketch is more challenging. Remember that the only dimensions that can be transferred directly to an isometric sketch are the three primary dimensions. These dimensions will follow the lines of the grid paper. When blocking in an isometric sketch, lay out the dimensions on construction lines that run parallel to the grid lines. Angled surfaces are created using construction lines that are nonisometric; that is, they do not run parallel to any of the grid lines and are drawn indirectly by connecting points marked on isometric construction lines.

Practice Exercise 2.2

Using isometric grid paper, sketch common, everyday objects. Some examples are given in Figure 2.31. Sketch objects with a variety of features. Some should require sketching isometric ellipses, while others should have angled surfaces that require nonisometric lines. Start with simpler forms that only contain isometric lines and work toward more complex

Trackball

Stapler

Screwdriver

Cassette Tape

Figure 2.31   Isometric Sketches of Common Objects

forms. Another approach is simply to leave out some of the details. You can capture the *essence* of the form by representing just its primary features. This is a common approach in creating ideation sketches.

The cost and availability of isometric grid paper can be a discouraging factor in using it to create lots of sketches. You can minimize the expense by using roll tracing paper over a sheet of grid paper. The two sheets can be held together with low-tack tape or put in a clipboard. With practice, you will find that grid paper is not needed and you can create sketches on the tracing paper alone.

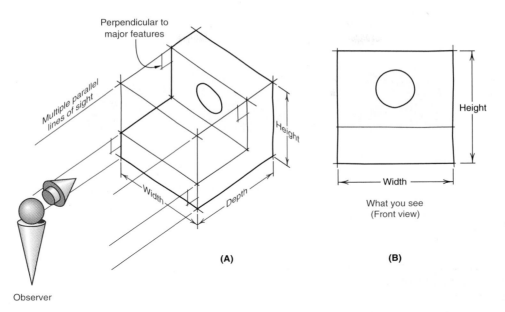

Perpendicular to
major features

Multiple parallel
lines of sight

Height

Width

Depth

Height

Width

Height

What you see
(Front view)

(A)

(B)

Observer

Figure 2.32 Multiview Drawings
Multiview drawings are classified as a parallel projection, because the lines of sight used to view the object are parallel. This method of viewing an object results in a single view, with only two of the three dimensions represented. Therefore, it takes multiple views to show all three dimensions.

2.4.4 Multiview Projections

Multiview drawings are based on parallel projection techniques and are used when there is a need to represent the features of an object more accurately than is possible with a single (pictorial) view. A multiview drawing is a collection of *flat* 2-D drawings that work together to give you an accurate representation of the overall object. With a pictorial drawing, all three dimensions of the object are represented in a single view. The disadvantage of this approach is that not all the features in all three dimensions can be shown with optimal clarity. In a multiview projection, however, each view concentrates on only two dimensions of the object, so particular features can be shown with a minimum of distortion. (Figure 2.32) Enough views are generated to capture all the important features of the object.

Given their advantages, why are multiview drawings not always used? For one thing, there are the multiple views, rather than a single view, to be drawn. These views must be coordinated with one another to represent the object properly. You have to carefully visualize the views as you sketch them, and so does the person viewing them.

Without training and experience, you might find it hard to interpret multiview drawings. The choice between multiviews and pictorials is often one of exact representation of features versus ease of sketching and viewing.

Orienting and Selecting the Front View   When creating a multiview drawing of a design, the selection and orientation of the front view is an important first step. The front view is chosen as the most descriptive of the object; for example, what would normally be thought of as the side of the car is chosen as the front view because it is the most descriptive. (Figure 2.33A) In addition, the object must be properly oriented in the view. The orientation of the object is based on its function. For example, for an automobile, the normal or operating position is on its wheels, rather than on its roof or bumpers. (Figure 2.33B)

Choosing the Views for a Multiview Drawing   Another way to understand the views of an object is to pick it up and turn it around. This may be hard to imagine with something like a car, but many of the objects you will be

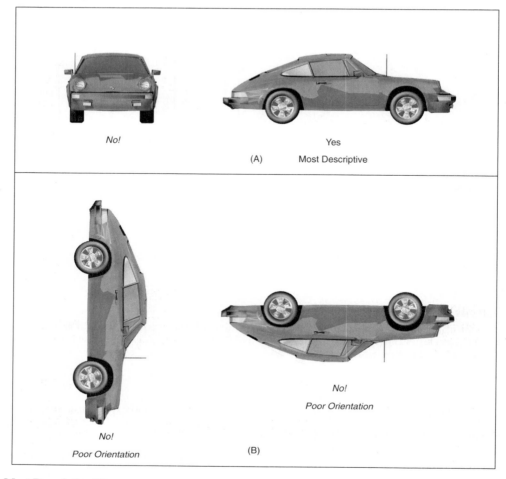

Figure 2.33   **Most Descriptive View**

Proper orientation and the most descriptive features of an object are important when establishing the front view for a multiview sketch. Objects should be positioned in their natural orientation; for this car, that position is on its wheels.

sketching are considerably smaller and can be held in the hand. Imagine picking up the object shown in Figure 2.34 and holding it so that you are looking at the front. Now, rotate the object so that you are looking at its top. Rotate it back to where you started and then rotate it so you are looking at its right side. There are an infinite number of intermediate views of the object between the points of rotation at which you stopped; for right now, however, consider only those views that are rotated 90 degrees from each other. These are considered *regular* or *principal views,* and each represents two primary dimensions of the object. If you continue rotating the object in 90-degree increments, you will see as many as six regular views. (Figure 2.35)

A multiview drawing should have the minimum number of views necessary to describe an object completely.

Normally, three views are all that are needed; however, the three views chosen must be the most descriptive ones. The most descriptive views are those that reveal the most information about the design, with the fewest features hidden from view.

For example, Figure 2.35 shows one isometric view and six orthographic views of a simple object. Of the six orthographic views, which are the most descriptive? Although all the views given reveal much about the size and shape of the object, some are more descriptive than others. The choice of views is determined as follows:

1. Identify the most descriptive or important features of the object.

2. Determine the views that best represent those features.

Figure 2.34   Visualizing a Multiview Drawing
By rotating a real object in your hand, you can simulate how a multiview drawing is created. A different principal view of the object is produced for every 90 degrees of rotation.

After deciding the most descriptive features of the part, choose the views that show these features. Part of this selection will involve determining which views do the best job of neither obscuring nor hiding other features. For example, the object in Figure 2.35 has four important features: the hole, the rounded top, the L-shaped profile, and the slot cut out of the base in front. There are only two views that show the hole and rounded top: the front and rear. Although both views show these features equally well, the front view is chosen over the rear view because it does not hide the slot cut in the base. The L-shaped profile is shown equally well in both the right and left side views, and they have an equal number of hidden features. Although either view can be used, convention favors choosing the right side view. The slot in the base is shown in both the top and bottom views. However, the top view has fewer hidden lines, so it is preferred over the bottom view for the sketch. For this object then, the preferred views are the front, top, and right side views.

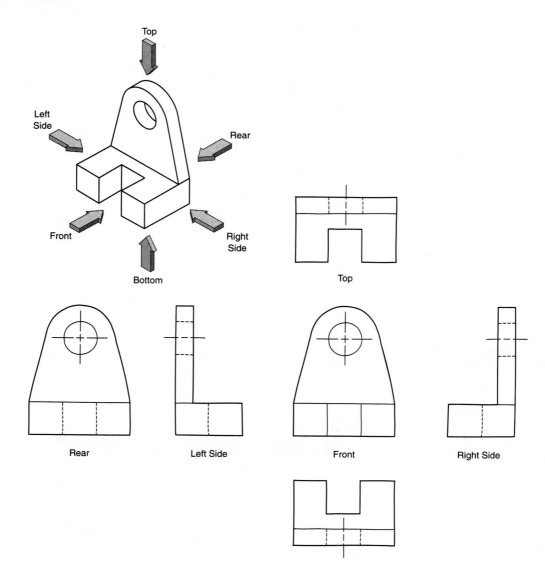

Figure 2.35 Six Principal Views

A multiview drawing of an object will produce six views, called regular or principal views. The object is viewed from six mutually perpendicular viewpoints. The six views are called front, top, bottom, rear, right side, and left side.

Practice Exercise 2.3

Find a number of small objects that can be picked up in your hand and rotated. Look at them carefully and identify their important features. Orient each object so that the most important features are seen. Make this view the front view. Before rotating them, try to imagine what the top and right side views will look like. Now rotate them and see if that is how they look. Do all of the important features show up in these three views? If not, what other front view can you select that would result in seeing all of the important features in the three views?

Next, look at some larger items you cannot pick up. For each object, move around it and look at it from different viewpoints. Try to imagine picking it up and rotating it. Is there any difference in how the object looks if you pick it up and rotate it rather than walk around it?

Figure 2.36 **Multiviews of Common Objects**
These multiview drawings of common objects show the front, top, and right side views. Hidden lines have been omitted for clarity.

Figure 2.36 shows some common objects represented in multiview drawings.

2.5 | MULTIVIEW SKETCHING TECHNIQUE

As with other types of drawings, there is a certain technique, or standard set of methods, followed when creating multiview sketches. The technique includes line conventions, proportioning, and methods for creating circles, arcs, and ellipses. A knowledge of the proper and effective technique will assist the beginner in creating legible multiview sketches.

2.5.1 Line Conventions

As in all engineering and technical drawings, multiview drawings and sketches require adherence to the proper use of the alphabet of lines. Figure 2.37 shows the alphabet of lines, sketched rather than drawn with drawing instruments, and includes the recommended pencil thicknesses. Figure 2.38 shows the application of each linetype on an engineering sketch.

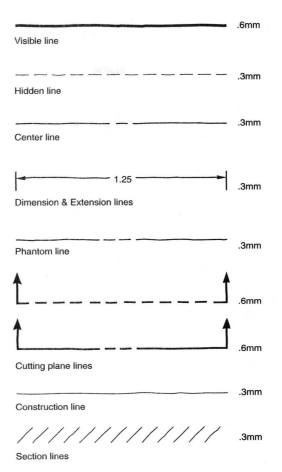

Visible line	.6mm
Hidden line	.3mm
Center line	.3mm
Dimension & Extension lines (1.25)	.3mm
Phantom line	.3mm
Cutting plane lines	.6mm
	.6mm
Construction line	.3mm
Section lines	.3mm

Figure 2.37 **Sketched Alphabet of Lines**
Standard engineering drawing practice requires the use of standard linetypes, which are called the alphabet of lines. The sizes show the recommended line thicknesses.

In engineering and technical drawing, it is important that hidden features be represented, so that the reader of the drawing can clearly understand the object. Many conventions related to hidden lines have been established over the years. Figure 2.39 shows the **hidden line conventions** that must be followed when creating technical sketches or instrument drawings unless a CAD system has limitations that make it difficult to comply with these requirements. These conventions are as follows:

- There should be no gap when a hidden line intersects a visible line. (Figure 2.39A)
- Corners on hidden lines should be joined. (Figure 2.39B)

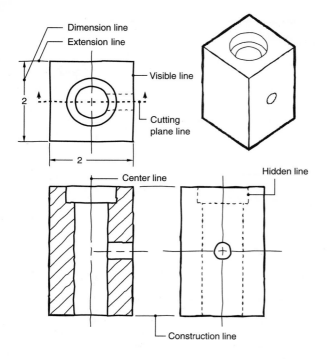

Figure 2.38
This engineering sketch has labels to identify some of the alphabet of lines.

- There should be a gap when a hidden line intersects either a visible corner or visible arc. (Figure 2.39C)
- Three (hidden) intersecting corners, found in holes that are drilled and that end inside the object (i.e., do not go all the way through the object), should be joined as shown in Figure 2.39D.
- At the bottom of the drilled hole, the lines indicating the tip (created by the drill, which has a pointed tip) are joined. (Figure 2.39E)
- Hidden arcs are started on the center line or the point of tangency. (Figure 2.39F)
- When a hidden line passes behind a visible line (i.e., does not intersect the visible line), do not put a hidden-line dash on the visible line. (Figure 2.39G)
- At the point where one hidden line crosses in front of another hidden line (indicating two hidden features, one closer to the visible view than the other), use a dash for the hidden line in front; that is, if the front hidden line is horizontal, use a horizontal dash at the point of crossing. (Figure 2.39H)

2.5.2 Precedence of Lines

It is fairly common in technical drawings to have two lines in a view coincide. When this happens, the conventional practice called the **precedence of lines** dictates the line-type to draw when two or more lines in a view overlap. (Figure 2.40)

For example, in Figure 2.41A a visible line in the top view coincides with the hidden lines for the hole. The precedence of lines requires that the visible lines be drawn and the hidden lines not be shown in the top view. Figure 2.33B shows an example of a hidden line that has precedence over a center line. Figure 2.41C is an example of a visible line having precedence over a center line. Notice that whenever a hidden or visible line has precedence over a center line, the center line is still drawn in the view by leaving a space and then extending it beyond the edge. (Figure 2.41D)

2.5.3 Conventional Practices for Circles and Arcs

Circles are used to represent holes and the bases of cones and cylinders. Arcs are used to represent portions of these elements, as well as rounded corners on objects. Whenever representing a hole, cylinder, or cone on a technical drawing, the conventional practice is to draw center lines, which are used to (1) locate the centers of circles and arcs; (2) represent the axis of cylinders, cones, and other curved surfaces; and, (3) represent lines of symmetry.

Figure 2.42 is a multiview drawing of a cylinder. In the top view, horizontal and vertical center lines are drawn to locate the center of the circle. In the front view, a very thin center line (.035 mm) is drawn to show the location of the cylinder's axis. The small dashes that cross at the center of the circle extend approximately 8 mm or ⅜″ from the edges of the object. The short segment of the center line is approximately 3 mm or ⅛″ long. The long segment can be from 20 mm to 40 mm, or ¾″ to 1-½″ long. For very long cylinders, the center line is drawn as a series of long and short segments.

Figure 2.43 shows some applications and the associated conventions for placing center lines. Notice that center lines are used in both the circular and horizontal views of a hole. When adding center lines to the circular view of a very small hole, a solid center line may be used rather than a dashed line, as shown in part C. Part D shows how center lines are used to locate the centers of holes around a bolt circle. Part E shows how center lines, along with phantom lines, are used to show a path of motion.

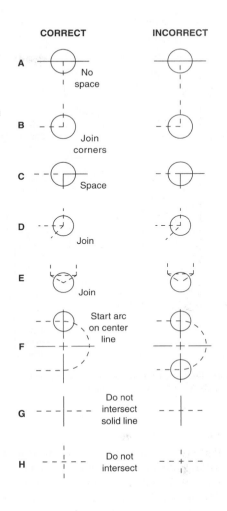

Figure 2.39   **Drawing Conventions for Hidden Lines**

2.6 | MULTIVIEW SKETCHES

Multiview drawings can have from one to three or more views of an object. However, multiview sketches rarely have more than three views.

Multiview sketches are important in that they provide more accurate geometric information than a pictorial sketch, without requiring the time that a formal multiview drawing would take. If dimensions are provided, they are usually only for a particular feature(s) and are often only approximations, since these sketches are used early in the design process before final specifications have been made.

As is the goal with all sketches, multiview sketches should be done quickly and clearly. Straightedges, such as triangles and T-squares, should be avoided since they will

VISIBLE LINE takes precedence over all other lines

HIDDEN LINE and CUTTING PLANE LINE take precedence over center lines

CENTER LINE does not have precedence

Figure 2.40   **Precedence of Lines**

The precedence of lines governs which lines are drawn when more than one line occupies the same position on a drawing. For example, a visible line has precedence over all other types of lines, and a hidden line and a cutting plane line have precedence over a center line.

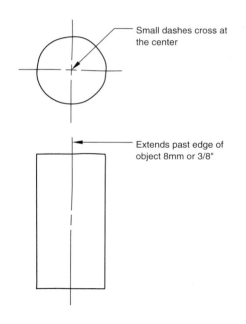

Figure 2.42   An Engineering Drawing of a Cylinder, Showing the Application of Center Lines

Figure 2.41   An Engineering Drawing Showing How the Precedence of Lines Is Applied

only slow you down and will compel you toward a level of finish that is inappropriate in sketches. In addition, you should draw only as many views as are necessary to show the features accurately. An initial analysis of the features should tell you if one, two, or three views are needed to clearly show the elements of interest.

2.6.1 One-View Sketches

The simplest multiview sketch represents just the front view of an object. Though not truly a *multiview,* it is still meant to represent only two dimensions of the object, which is the basic concept of multiview drawings. This sketch can be produced using the techniques shown in Figure 2.20.

2.6.2 Two-View Sketches

Occasionally, an object can be completely described using only two views. As an example, Figure 2.43 shows a symmetrical part that can be described using two views. If the

front view is as shown in the pictorial, the top and side views would be the same. Nothing would be gained by drawing both the top and side views, so only one of these views is sketched.

Creating a Two-View Sketch

Figure 2.44 and the following steps describe how to create a two-view sketch.

Step 1.   In the front view, block in the squares with sides equal to the diameters of the circles. Since both the front and right side views show the height dimension, construction lines can be used to project the height of the squares onto the right side view. Block in the rectangles representing the details for the side view.

Step 2.   Using the squares and center lines as guides, sketch the circles for each hole and the cylinder, in the front view. Using the construction lines as guides, sketch the hidden lines for the holes, in the side view.

Step 3.   Darken all visible, center, and hidden lines.

2.6.3 Three-View Sketches

When an object is more complex, three views are needed. The object used in Figure 2.45 was chosen because it has

Figure 2.43 Center Line Conventions
These engineering drawings show various applications for center lines. Study each example to learn how center lines are used.

circles, arcs, center lines, and hidden lines.

Front Right Side

Height

(A) CORRECT! (B) INCORRECT! (C) INCORRECT! BOTTOM (D)

?

R. SIDE FRONT

Figure 2.46 View Alignment

A three-view sketch must be laid out properly, following standard practices. The width dimensions on the front and top views are aligned vertically, and the height dimensions on the front and side views are aligned horizontally.

view goes directly above and is aligned with the front view, and the right side view goes directly to the right of and is aligned with the front view. Do not rearrange the top, front, or right side views, or substitute other regular views in their places.

2.7 | PERSPECTIVE PROJECTION

Perspective projection is the projection method that represents three-dimensional objects on two-dimensional media in a manner closest to how we perceive the objects with our eyes. If you were to take a photograph, lay tracing paper over it, and sketch the objects in it, the result would look like a perspective projection. (Figure 2.47) Like the other pictorial sketches, all three dimensions of the object are presented in a single image.

If you were to stand in the middle of a straight, flat road and look toward the horizon (Figure 2.48), the road would appear to narrow to a single point, called the vanishing point. Even though the road appears to go to a point, in reality it does not. You know that the edges of the road are parallel to each other and, as you travel down the road, the portion of the road that looked so small before will be full size. Objects, whether they are portions of a road, cars, or billboards, look smaller as they get farther away. Also, parallel lines—such as the two edges of the road—will appear to *converge* (come together) as they recede from you. Through the use of construction lines in your perspective sketch, you can control the convergence of the parallel edges of an object, as well as the proportional sizes of objects as they recede in the distance. (Figure 2.49)

Figure 2.47 Convergence in Photographs
This photograph shows an object in perspective, with major lines of convergence overdrawn.

Figure 2.49 shows the labels for the important elements of a perspective sketch. Most important is the **horizon line (HL)**, which is an imaginary line in the distance, where objects are infinitely small and parallel lines converge. The point on the horizon line where parallel lines converge is called the **vanishing point (VP)**. Where the portion of the object closest to the observer rests on the ground plane is called the **ground line (GL)**, as shown in Figures 2.49 and 2.50.

The horizon line shown in Figures 2.49 and 2.50 also represents the observer's eye level. The relationship of the ground line to the horizon line reflects the height of the observer relative to the object. Figure 2.51 shows the different types of views created by raising or lowering the ground line, and each view type has a specific use, as shown in Table 2.1.

B.C.

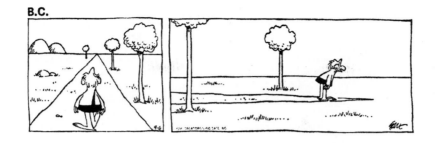

Figure 2.48

Humans view their environment in perspective, where lines appear to converge to a single point. Even though it looks as though the edges of a road converge on the horizon, we know they don't. By permission of Johnny Hart and Creators Syndicate, Inc.

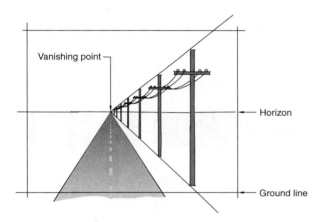

Figure 2.49   Principal Elements in Perspective Sketches

Principal elements of a perspective sketch include the horizon, vanishing point, and ground line. All elements in a perspective sketch are drawn to a single vanishing point.

The human's eye view is the most commonly used for sketching everyday objects. The ground's eye is useful for one- to three-story buildings, and the worm's eye view is used for taller structures. (Figure 2.52)

Figures 2.48 through 2.51 show perspective projections with only one vanishing point. In Figures 2.50 and 2.51, the parallel edges of only one of the object's three dimensions converge to the vanishing point. If you want more of the dimensions to converge, more vanishing points are needed. Perspective views are classified according to the number of vanishing points used to create the drawing.

Figure 2.53 shows one-, two-, and three-point perspective sketches. Although three-point perspective is the most

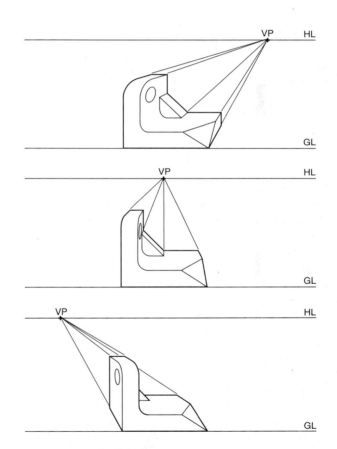

Figure 2.50   Moving the Vanishing Point Right to Left

The view of the object can be dramatically changed by moving the vanishing point along the horizon line. Points to the right of center will reveal details about the right side of the object; points to the left of center will reveal details about the left side of the object.

Table 2.1   Relationship between the Ground Line and Horizon Line

View Name	Relationship	Effect
Bird's eye view	Ground line well below the horizon line.	Observer is much taller than the object.
Human's eye view	Ground line six feet below the horizon line.	Observer is somewhat taller than the object.
Ground's eye view	Ground line at the same level as the horizon line.	Observer is the same height as the object.
Worm's eye view	Ground line well above the horizon line.	Observer is much shorter than the object.

Bird's Eye View—Ground Line Below Horizon Line

Human's Eye View—Ground Line 6' Below Horizon

Ground's Eye View—Ground Line on the Same Level as the Horizon Line

Worm's Eye View—Ground Line Above the Horizon Line

Figure 2.51   Moving the Vanishing Point Vertically
The view of the object can be dramatically changed by moving the ground line relative to the horizon line.

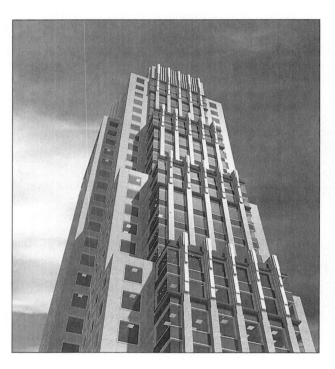

Figure 2.52   A Worm's Eye Perspective View of a Building, Created with CAD
This view simulates what the building would look like when viewed looking up from the ground. Courtesy of Michael Sechman and Associates, Oakland, California.

realistic, one- and two-point perspectives are usually adequate and are simpler to sketch. By varying the number and positions of the vanishing points, and the position of the ground line relative to the horizon line, it is possible to create virtually any perspective view of an object.

2.7.1 One-Point Perspective Sketch

One-point perspective sketches are used to quickly create a pictorial representation of an object or structure.

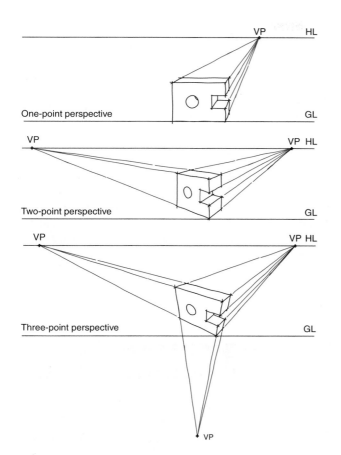

Figure 2.53 One-, Two-, and Three-Point Perspective Sketches of the Same Object

Within image 1 the labels read:
One-point perspective
Two-point perspective
Three-point perspective

Creating a One-Point Perspective Sketch

The following steps describe how to create a one-point perspective sketch of the guide block shown in Figure 2.54.

Step 1. For a one-point perspective sketch, begin by determining the type of view you want (human's eye, ground's eye, etc.); then place the corresponding horizon and ground lines on the paper.

Step 2. Establish the relationship of the vanishing point to the object, such as to the right of the object. With the vanishing point marked, box in a front view of the object with construction lines, marking the height and width dimensions of the features of the object.

Step 3. From three corners of the bounding box (points 1, 2, and 3), draw converging lines to the vanishing point. The fourth corner would be hidden, so it is not drawn. Based on the depth of the object, determine where the back of the object would be on the converging lines, and draw a horizontal and a vertical line to mark the back of the object.

Figure 2.54 Constructing a One-Point Perspective Sketch, Using Five Steps

Step 4. A bounding box in all three dimensions is now established for the object. Continue refining the object by sketching more bounding boxes representing other features. Remember that bounding boxes drawn on the front and back faces of the object will be square, while those placed on the top and sides will not. On the right side, the vertical edges will all be parallel, while on the top the horizontal edges will be parallel. Those edges going back in depth on either face will converge at the vanishing point.

Step 5. Finally, sketch the dark lines representing the object, paying close attention to those edges that converge.

2.8 | LETTERING

All engineering and technical drawings, including sketches, contain some notes or dimensions. All text on a drawing must be neat and legible, so it can be easily read on both the original drawing and a reproduction, such as a blueprint or photocopy. Although precise lettering is not required, legibility is very important.

Until the invention of printing by Johann Gutenburg in the 15th century, all text was hand lettered, using a very personalized style. With the invention of printing, text styles became more standardized. Although some early technical drawings were embellished with personalized text, current ANSI standards suggest the use of single-stroke Gothic characters, for maximum readability. (Figure 2.55)

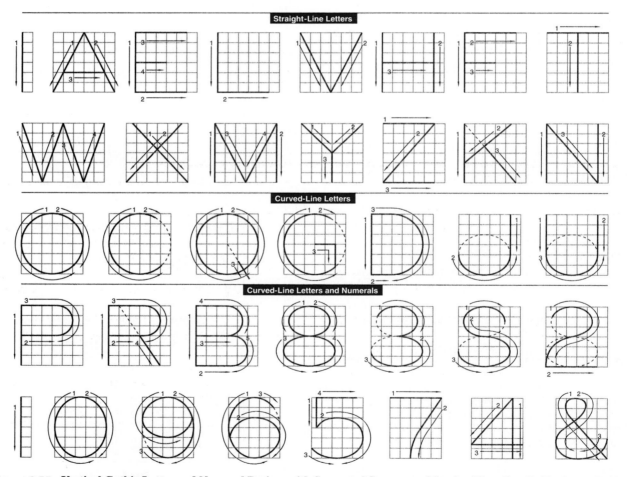

Figure 2.55 Vertical Gothic Letter and Numeral Design, with Suggested Sequence of Strokes That Can Be Used as a Guide for Hand Lettering a Technical Drawing

Questions for Review

1. Define and describe the uses for technical sketching.
2. Define an ideation sketch, and explain how it differs from a document sketch.
3. List the four types of sketches, grouped by projection methods. Sketch an example of each type.
4. Describe the major differences between parallel and perspective projection.
5. Define multiview drawing, and make a sketch of one.
6. Define principal view.
7. Describe the precedence of lines.
8. List the two important uses for text on a drawing.

Further Reading

Edwards, B. *Drawing on the Right Side of the Brain.* New York: St. Martin's Press, 1979.

Hanks, K., and L. Belliston. *Draw!* Los Altos, CA: William Kaufmann, 1977.

Hanks, K., L. Belliston, and D. Edwards. *Design Yourself!* Los Altos, CA: William Kaufmann, 1978.

Knowlton, K. W. *Technical Freehand Drawing and Sketching.* New York: McGraw-Hill, 1977.

Martin, L. C. *Design Graphics.* New York: Macmillan, 1968.

Problems

Use the grided problem sheets provided.

Hints for Isometric Sketching

- Identify the major features and overall dimensions of the object.
- Use clean, crisp strokes.
- Do not use straightedges or scales when sketching.
- Start by drawing a bounding box, using construction lines.
- Only measure dimensions along the primary axes.
- Do not directly transfer angles from a multiview to a pictorial.
- Use light construction lines to locate vertices and edges.
- Sketch faces roughly in this order:

  1. Normal faces on the perimeter of the bounding box.
  2. Normal faces in the interior of the bounding box.
  3. Inclined faces.
  4. Oblique faces.

- Darken all object lines.

Hints for Multiview Sketching

- Identify the major features and overall dimensions of the object.
- Use clean, crisp strokes.
- Do not use straightedges or scales when sketching.
- Start by drawing bounding boxes and a miter line, using construction lines.
- Align the views.
- Use light construction lines to locate vertices and edges.
- Only measure dimensions along the primary axes.
- Map inclined and oblique faces between all three views.
- Follow the precedence of lines.
- Doublecheck to make sure there are no missing hidden or center lines.
- Darken all visible, hidden, and center lines.

2.1 Use upside-down sketching to create the word sketch as shown in Figure 2.56.

2.2 Use upside-down sketching to create the table shown in Figure 2.57.

2.3 Use upside-down sketching to create the chair shown in Figure 2.58.

2.4 Use contour sketching to create the series of cubes and cylinders shown in Figure 2.59.

2.5 Use contour sketching to create the overlapping shapes shown in Figure 2.60.

2.6 Use contour sketching to create the optical illusions shown in Figure 2.61.

2.7 Create a negative space sketch of the paper clips shown in Figure 2.62.

2.8 Using visible, center, and hidden line styles, practice sketching the following types of lines/features using size A or A4 plain or grid paper:

- Straight lines.
- 90- and 180-degree arcs.
- Circles.
- Ellipses.

Figure 2.58   **Problem 2.3 Upside-Down Sketch of a Chair**

HƆTƎ⋊S

Figure 2.56   **Problem 2.1 Upside-Down Sketch of the Word SKETCH**

Figure 2.57   **Problem 2.2 Upside-Down Sketch of a Table**

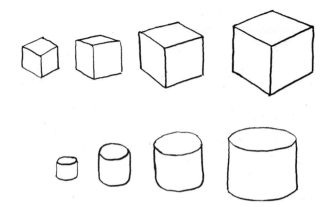

Figure 2.59   **Problem 2.4 Contour Sketching of Cubes and Cylinders**

2.9 Refer to Figures 2.63A–D, Multiview Sketching Problems. Sketch freehand multiview representations of the given objects using either size A or A4 plain or grid paper.

2.10 Refer to Figures 2.64A–C, Pictorial Sketching Problems. Sketch freehand pictorials of the given objects using either size A or A4 plain or grid paper.

2.11 Draw a pictorial sketch of one of the multiviews in Problem 2.10. Pass the pictorial sketch to another student in the class, but do not specify which multiview you used. Have that student draw a multiview from your sketch. Compare that with the original multiview. Do they look the same? Reverse the process by starting with one of the objects in Problem 2.9.

2.12 Using one object from either Problem 2.9 or 2.10, sketch isometric pictorials from three different viewpoints, on size A or A4 plain or grid paper.

2.13 Refer to Figures 2.63 and 2.64. Create one-point perspective sketches.

Figure 2.60 **Problem 2.5 Contour Sketch of Overlapping Shapes**

Figure 2.62 **Problem 2.7 Negative Space Sketching**

Figure 2.61 **Problem 2.6 Contour Sketching of Optical Illusions**

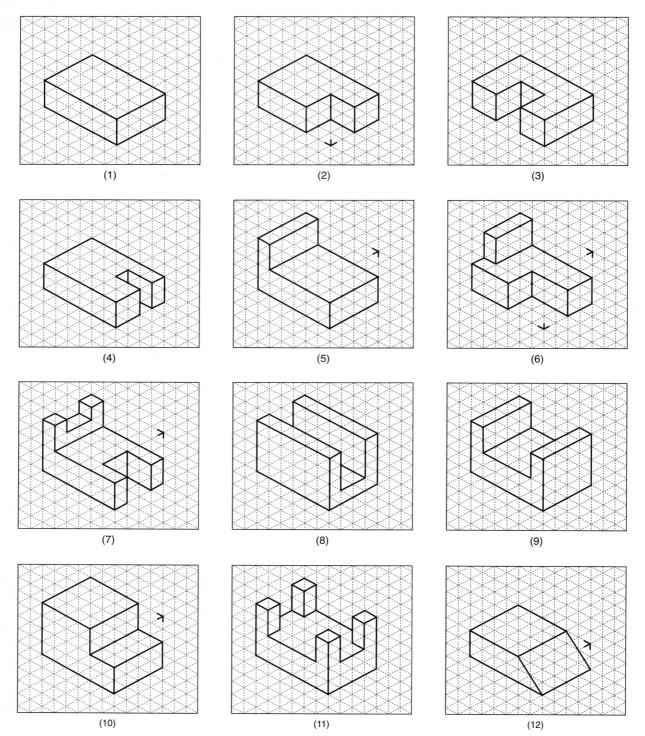

(1)   (2)   (3)

(4)   (5)   (6)

(7)   (8)   (9)

(10)   (11)   (12)

Figure 2.63A   Problem 2.9 Multiview Sketching Problems

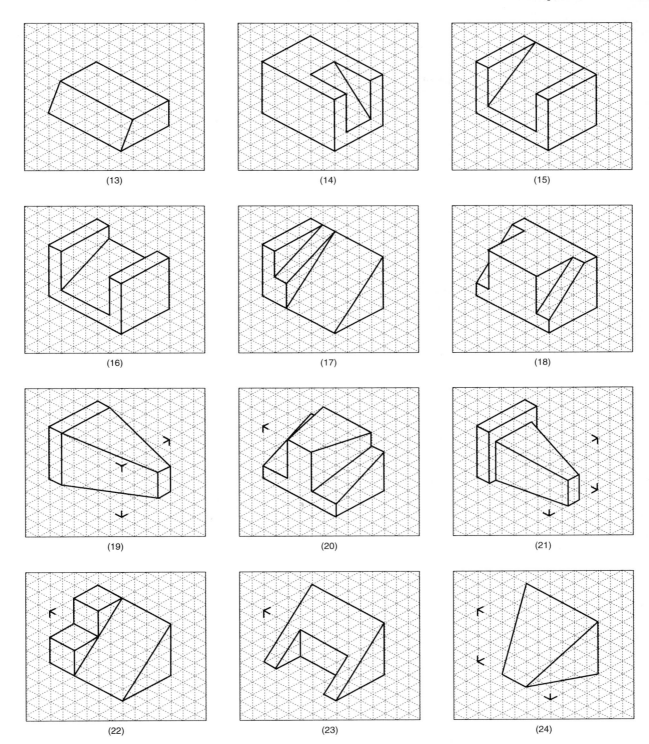

(13) (14) (15)

(16) (17) (18)

(19) (20) (21)

(22) (23) (24)

Figure 2.63B **Problem 2.9 Multiview Sketching Problems**

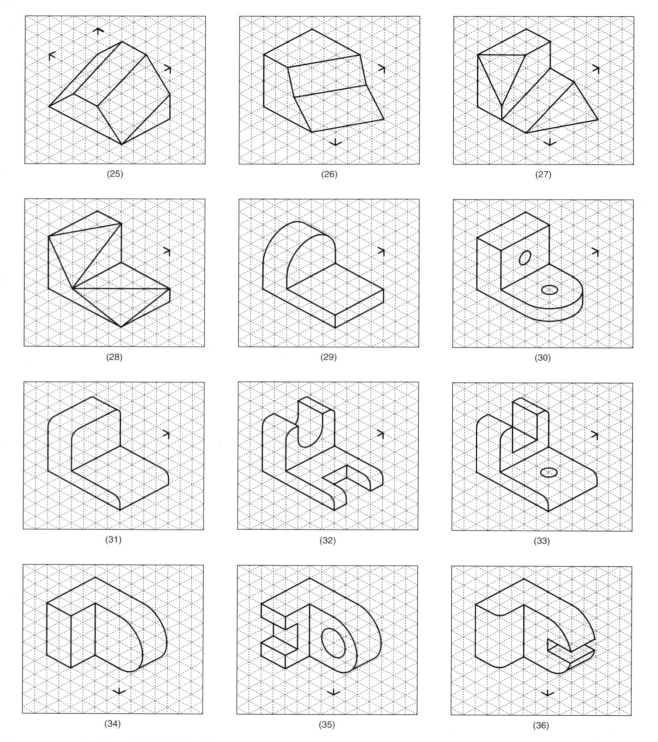

(25)   (26)   (27)

(28)   (29)   (30)

(31)   (32)   (33)

(34)   (35)   (36)

Figure 2.63C   Problem 2.9 Multiview Sketching Problems
Assume all holes to be through.

(37)　　(38)　　(39)

(40)　　(41)　　(42)

(43)　　(44)　　(45)

(46)　　(47)　　(48)

Figure 2.63D   Problem 2.9 Multiview Sketching Problems
Assume all holes to be through.

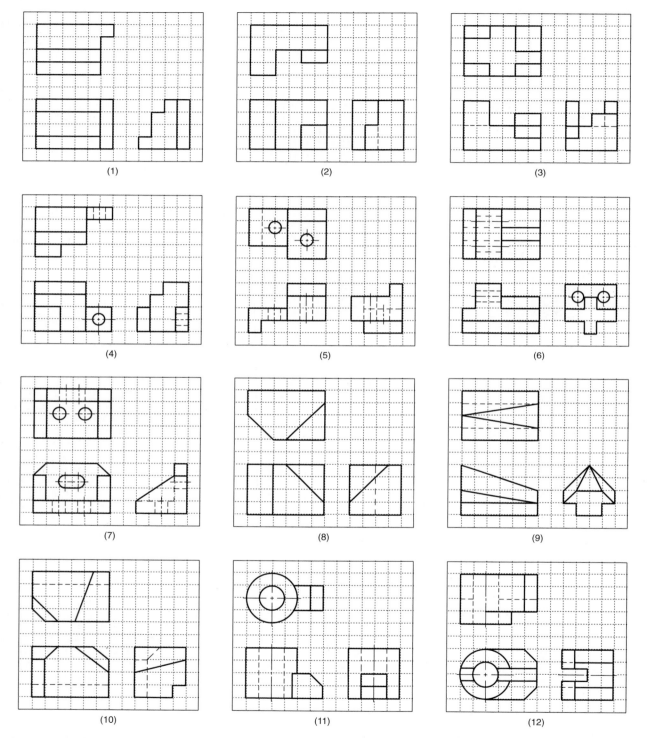

Figure 2.64A Problem 2.10 Pictorial Sketching Problems

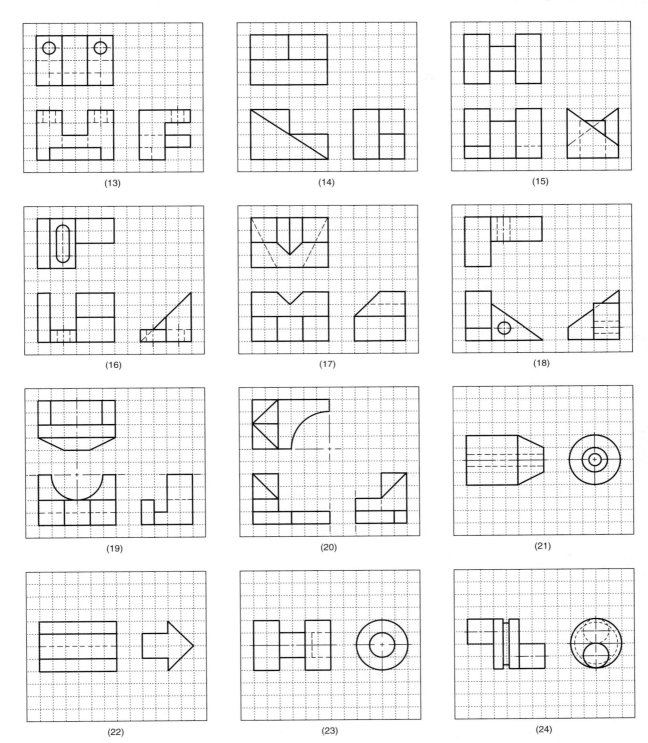

Figure 2.64B **Problem 2.10 Pictorial Sketching Problems**

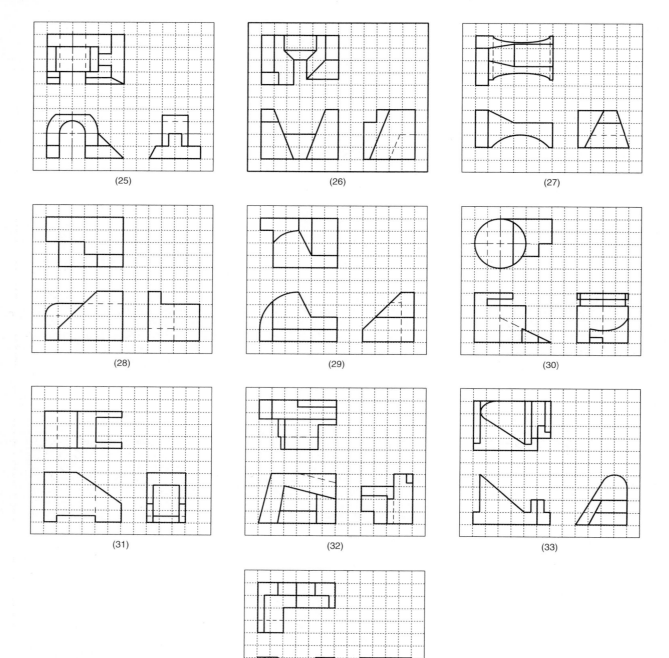

(25)   (26)   (27)

(28)   (29)   (30)

(31)   (32)   (33)

(34)

Figure 2.64C   Problem 2.10 Pictorial Sketching Problems

Problems 2.1 through 2.5

Refer to Figures 2.56 through 2.62. Create those sketches assigned by your instructor.

Sketch Number: _____

Name:_____

Div/Sec: _____

Date:_____

Problems 2.1 through 2.5

Refer to Figures 2.56 through 2.62. Create those sketches assigned by your instructor.

Sketch Number: _____

Name:_____

Div/Sec: _____

Date:_____

Problem 2.8

Using visible, center, and hidden line styles, practice sketching straight lines, 90- and 180-degree arcs, circles, and ellipses.

Sketch Number: _____

Name:_____

Div/Sec: _____

Date:_____

Problem 2.9

Refer to Figure 2.63. Sketch multiviews of the pictorial views assigned. Divide the long dimension of the paper with a thick dark line.

Sketch Number: _____

Name:_____

Div/Sec: _____

Date:_____

Problem 2.9

Refer to Figure 2.63. Sketch multiviews of the pictorial views assigned. Divide the long dimension of the paper with a thick dark line.

Sketch Number: _____

Name: _____

Div/Sec: _____

Date: _____

Problem 2.9

Refer to Figure 2.63. Sketch multiviews of the pictorial views assigned. Divide the long dimension of the paper with a thick dark line.

Sketch Number: _____

Name: _____

Div/Sec: _____

Date: _____

Problem 2.10

Using one object from either Problem 2.9 or 2.10, sketch isometric pictorials from three different viewpoints.

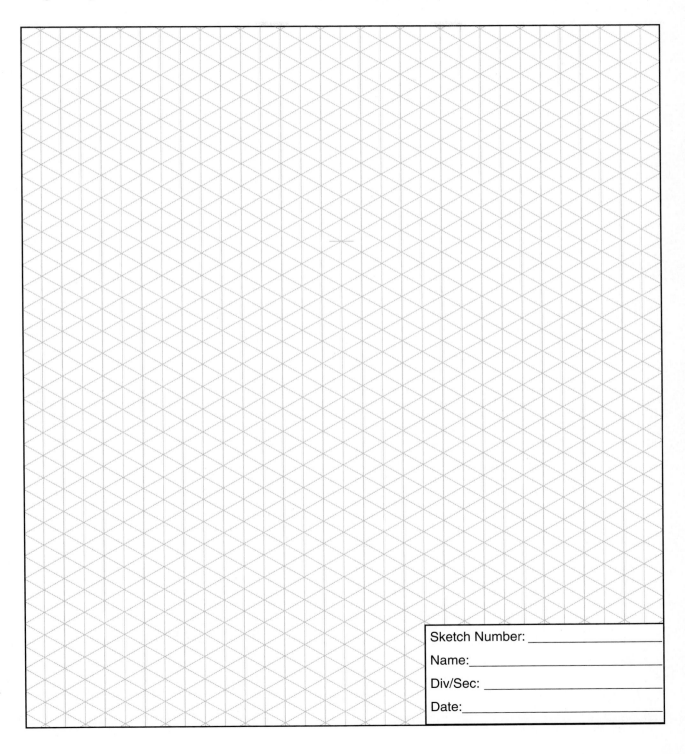

Sketch Number: _____

Name:_____

Div/Sec: _____

Date:_____

Problem 2.10

Using one object from either Problem 2.9 or 2.10, sketch isometric pictorials from three different viewpoints.

Sketch Number: _____

Name:_____

Div/Sec: _____

Date:_____

Problem 2.10

Using one object from either Problem 2.9 or 2.10, sketch isometric pictorials from three different viewpoints.

Sketch Number: _____

Name: _____

Div/Sec: _____

Date: _____

Problem 2.12

Using one object from either Problem 2.9 or 2.10, sketch isometric pictorials from three different viewpoints.

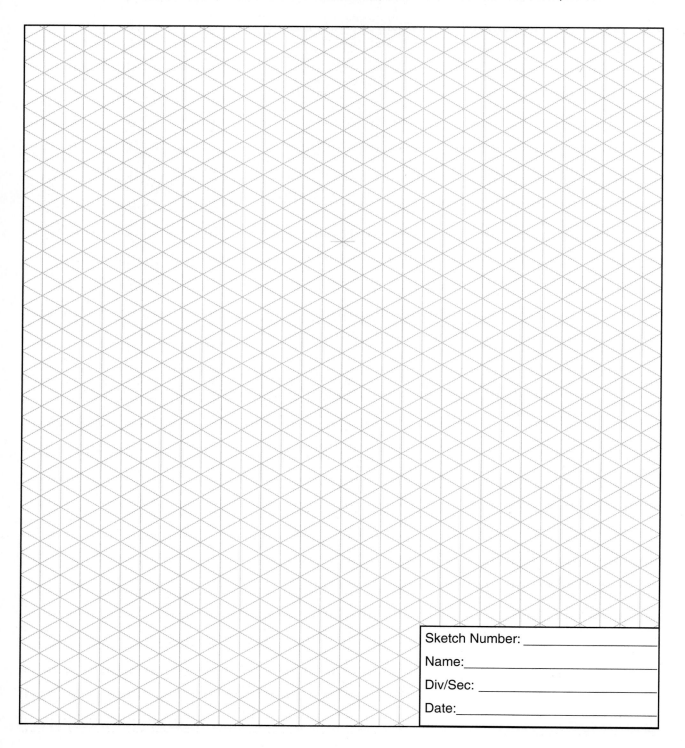

Sketch Number: _____

Name:_____

Div/Sec: _____

Date:_____

Problem 2.12

Using one object from either Problem 2.9 or 2.10, sketch isometric pictorials from three different viewpoints.

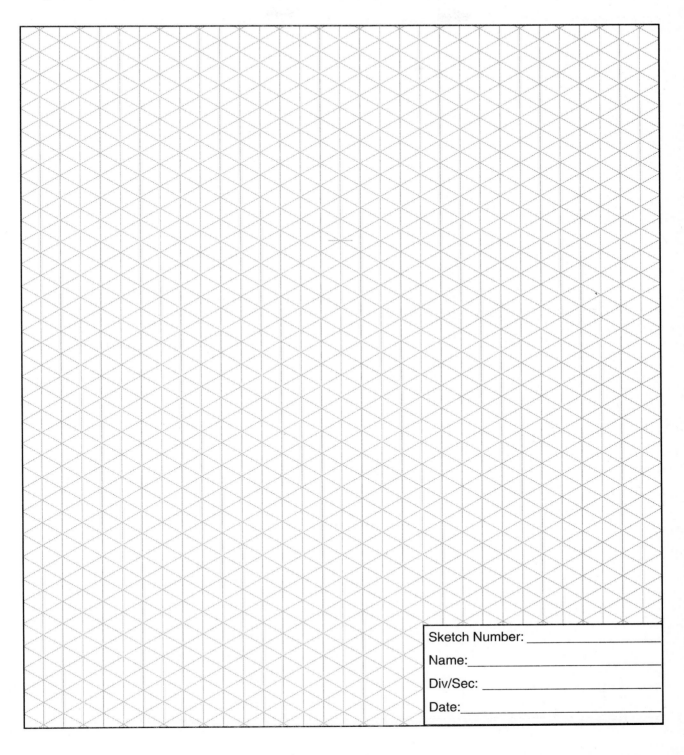

Sketch Number: _____

Name:_____

Div/Sec: _____

Date:_____

Problem 2.12

Using one object from either Problem 2.9 or 2.10, sketch isometric pictorials from three different viewpoints.

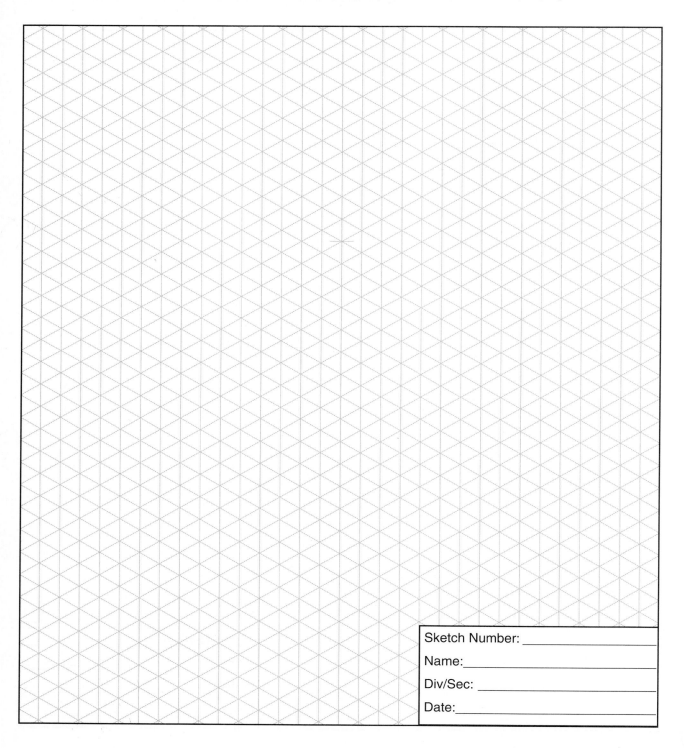

Sketch Number: _____

Name:_____

Div/Sec: _____

Date:_____

Problem 2.13

Refer to Figures 2.63 through 2.64. Create those sketches assigned by your instructor.

Sketch Number: _____

Name:_____

Div/Sec: _____

Date:_____

Section and Auxiliary Views   3

OBJECTIVES

After completing this chapter, you will be able to:

1. Apply the concept of cutting planes to create section views.
2. Represent cutting plane lines and section lines, using conventional practices.
3. Create full, half, offset, removed, revolved, broken-out, auxiliary, and assembly section views, using conventional practices.
4. Create conventional breaks for different materials and cross sections.
5. Represent ribs, webs, and thin features in section, using conventional practices.
6. Represent aligned sectioned features, using conventional practices.
7. Create auxiliary views.

3.1 | SECTIONING BASICS

Section views are an important aspect of design and documentation, and are used to improve clarity and reveal interior features of parts and structures. (Figure 3.1) Section views are also used in the ideation and refinement stages of engineering design to improve the communications and problem-solving processes. **Sectional drawings** are multiview technical drawings that contain special views of a part or parts, views that reveal interior features. A primary reason for creating a section view is

Figure 3.1

A typical multiview technical drawing shows the right side view in full section and removed section details.

Normal multiview drawing   Section view drawing

Figure 3.2   Section View Reveals Hidden Features
A section view will typically reveal hidden features so that the object is more easily visualized.

the elimination of hidden lines so that a drawing can be more easily understood or visualized. Figure 3.2 shows a regular multiview drawing and a sectioned multiview drawing of the same part in the front view; the hidden features can be seen after sectioning.

Traditional section views are based on the use of an imaginary cutting plane that cuts through the object to reveal interior features. (Figure 3.3) This imaginary cutting plane is controlled by the designer and can (1) go completely through the object (full section), (2) go halfway through the object (half section), (3) be bent to go through features that are not aligned (offset section), or (4) go through part of the object (broken-out section).

An important reason for using section views is to reduce the number of hidden lines in a drawing. A section view reveals hidden features without the use of hidden lines. (Figure 3.4) Adding hidden lines to a section view complicates the drawing, defeating the purpose of using a section. There are times, however, when a minimum number of hidden lines are needed to represent features other than the primary one shown by the section. (Figure 3.5)

3.1.1 Visualization of Section Views

Figure 3.6 is a multiview drawing of a part that is difficult to visualize in its three-dimensional form because of the many hidden lines. A section view is created by passing an imaginary cutting plane vertically through the center of the part. Figure 3.7 is an isometric representation of the part after it is sectioned. This isometric section view shows the interior features of the part more clearly and is an aid for visualizing the 3-D form. The corners of the isometric section view are numbered so that they can be compared with the orthographic section view.

In Figure 3.6, the cutting plane arrows in the top view point up to represent the direction of sight for producing a front view in full section. The direction of the arrow can also be thought of as pointing toward the half of the object being kept. The front half of the object is "removed" to reveal the interior features of the part.

All the surfaces touched by the cutting plane are marked with section lines. Because all the surfaces are the same part, the section lines are identical and are drawn in the same direction. The center line is added to the counterbored hole to complete the section view.

Line of sight

Imaginary cutting plane

Figure 3.3 Cutting Planes
Imaginary cutting planes used to create section views are passed through the object to reveal interior features.

Imaginary cutting plane

Hidden lines
(not shown in section view)

Edge view of cutting plane

Normal orthogonal
view

Section view

Figure 3.4 Treatment of Hidden Lines
Normally, hidden lines are omitted from section views.

Optional use
of a hidden line

Figure 3.5 Optional Use of Hidden Lines
Hidden lines can be shown in section views usually to eliminate
the need for another view.

Arrows in wrong direction: arrows should show the line of sight necessary for section view

No!

Correct cutting plane line

Incorrect cutting plane line

Figure 3.6　Placement of Cutting Plane Lines
The cutting plane line is placed in the view where the cutting plane appears on edge.

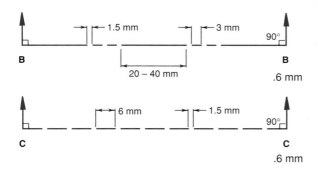

Figure 3.7　Standard Cutting Plane Linestyles
Standard cutting plane linestyles are thick lines terminated with arrows.

3.2 | CUTTING PLANE LINES

Cutting plane lines, which show where the cutting plane passes through the object, represent the *edge view* of the cutting plane and are drawn in the view(s) adjacent to the section view. In Figure 3.6, the cutting plane line is drawn in the top view, which is adjacent to the sectioned front view. Cutting plane lines are *thick* (0.6 mm or 0.032 inch) dashed lines that extend past the edge of the object ¼″ or 6 mm and have line segments at each end drawn at 90 degrees and terminated with arrows. The arrows represent the direction of the line of sight for the section view, and they point away from the sectioned view. Two types of lines are acceptable for cutting plane lines in multiview drawings, as shown in Figure 3.7.

3.3 | SECTION LINE PRACTICES

Section lines or **cross-hatch lines** are added to a section view to indicate the surfaces that are cut by the imaginary cutting plane. Different section line symbols can be used to represent various types of materials. However, there are so many different materials used in design that the general symbol (i.e., the one used for cast iron) may be used for most purposes on technical drawings. The actual type of material required is then noted in the title block or parts list or entered as a note on the drawing. The angle at which section lines are drawn is usually 45 degrees to the horizontal, but this can be changed for adjacent parts shown in the same section. Also, the spacing between section lines is uniform on a section view.

3.3.1 Material Symbols

The type of section line used to represent a surface varies according to the type of material. However, the **general-purpose section line** symbol used in most section view drawings is that of *cast iron.* Figure 3.8 shows some of the ANSI standard section line symbols used for only a few materials; there are literally hundreds of different materials used in design. Occasionally, a general section line symbol is used to represent a group of materials, such as steel. The specific type of steel to be used will be indicated in the title block or parts list. Occasionally, with assembly section views, material symbols are used to identify different parts of the assembly.

3.3.2 Drawing Techniques

The general-purpose cast iron section line is drawn at a 45-degree angle and spaced ¹⁄₁₆ inch (1.5 mm) to

Figure 3.8 ANSI Standard Section Lines for Various Materials

⅛ inch (3 mm) or more, depending on the size of the drawing.

Section lines should not run parallel or perpendicular to the visible outline. (Figures 3.9A and B) If the visible outline to be sectioned is drawn at a 45-degree angle, the section lines are drawn at a different angle, such as 30 degrees. (Figure 3.9C)

Avoid placing dimensions or notes within the section lined areas. (Figure 3.10) If the dimension or note must be placed within the sectioned area, omit the section lines in the area of the note, as shown in Figures 3.10B and C.

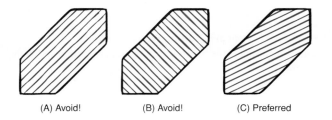

(A) Avoid!   (B) Avoid!   (C) Preferred

Figure 3.9 Section Lined Placement

Avoid placing section lines parallel or perpendicular to visible lines.

(A) Avoid!   (B) Preferred   (C) Preferred

Figure 3.10   Notes in Section Lined Areas
Section lines are omitted around notes and dimensions.

3.4 | SECTION VIEW TYPES

There are many different types of section views used on technical drawings:

Full section.
Half section.
Broken-out section.
Revolved section.
Removed section.
Offset section.
Assembly section.
Auxiliary section.

Up to this point, only full sections have been shown in the figures. The selection of the type of section view to be used is based upon which one most clearly and concisely represents the features of interest. For example, there are times when it may be more appropriate to use a half section view, for symmetrical objects, or a broken-out section, for small areas. The various types of section views are described in the following paragraphs.

3.4.1 Full Sections

A **full section** view is made by passing the imaginary cutting plane completely through the object, as shown in Figure 3.11A. Figure 3.11B shows the orthographic views of the object, and Figure 3.11C shows the full section view. All the hidden features intersected by the cutting plane are represented by visible lines in the section view. Surfaces touched by the cutting plane have section lines drawn at a 45-degree angle to the horizontal. Hidden lines are omitted in all section views unless they must be used to provide a clear understanding of the object.

The top view of the section drawing shows the cutting plane line, with arrows pointing in the direction of the line of sight necessary to view the sectioned half of the object. In a multiview drawing, a full section view is placed in the same position that an unsectioned view would normally occupy; that is, a front section view would replace the traditional front view.

3.4.2 Half Sections

Half sections are created by passing an imaginary cutting plane only *halfway* through an object. Figure 3.12A shows the cutting plane passing halfway through an object and *one quarter* of the object being removed. Figure 3.12B shows the normal orthographic views before sectioning, and Figure 3.12C shows the orthographic views after sectioning. Hidden lines are omitted on both halves of the section view. Hidden lines may be added to the unsectioned half for dimensioning or for clarity.

External features of the part are drawn on the unsectioned half of the view. A center line, not an object line, is

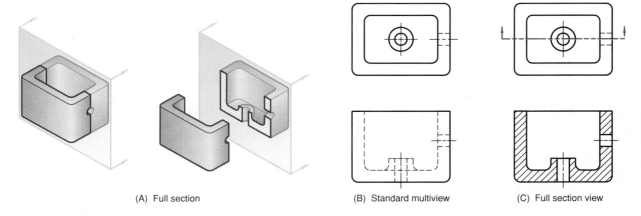

(A) Full section   (B) Standard multiview   (C) Full section view

Figure 3.11   Full Section View
A full section view is created by passing a cutting plane fully through the object.

(A) Half section   (B) Multiview   (C) Half section view

Figure 3.12 Half Section
A half section view is created by passing a cutting plane halfway through the object.

(A) Broken-out section   (B) Multiview   (C) Broken-out section view

Figure 3.13 Broken-Out Section
A broken-out section view is created by breaking off part of the object to reveal interior features.

used to separate the sectioned half from the unsectioned half of the view. The cutting plane line shown in the top view of Figure 3.12C is bent at 90 degrees, and one arrow is drawn to represent the line of sight needed to create the front view in section.

Half section views are used most often on parts that are symmetrical, such as cylinders. Also, half sections are sometimes used in assembly drawings when external features must be shown.

3.4.3 Broken-Out Sections

A **broken-out section** is used when only a portion of the object needs to be sectioned. Figure 3.13A shows a part with a portion removed or broken away. A broken-out section is used instead of a half or full section view to save time.

A break line separates the sectioned portion from the unsectioned portion of the view. A break line is drawn freehand to represent the jagged edge of the break. No cutting plane line is drawn. Hidden lines may be omitted from the unsectioned part of the view unless necessary for clarity, as shown in Figure 3.13C.

3.4.4 Revolved Sections

A **revolved section** is made by revolving the cross-section view 90 degrees about an axis of revolution and superimposing the section view on the orthographic view, as shown in Figure 3.14A. Visible lines adjacent to the revolved view can either be drawn or broken out using conventional breaks, as shown in Figure 3.14B. When the revolved view is superimposed on the part, the original lines of the part behind the section are deleted. The cross section is drawn true shape and size, not distorted to fit the view. The axis of revolution is shown on the revolved view as a center line.

A revolved section is used to represent the cross section of a bar, handle, spoke, web, aircraft wing, or other elongated feature. Revolved sections are useful for describing a cross section without having to draw another view. In addition, these sections are especially helpful when a cross section varies or the shape of the part is not apparent from the given orthographic views

3.4.5 Removed Sections

Removed section views do not follow standard view alignments as practiced in multiview drawings. Removed sections are made in a manner similar to revolved sections by passing an imaginary cutting plane perpendicular to a part then revolving the cross section 90 degrees. However, the cross section is then drawn adjacent to the orthographic view, not on it. (Figure 3.15) Removed sections are used when there is not enough room on the orthographic view for a revolved section.

Removed sections are used to show the contours of complicated shapes, such as the wings and fuselage of an airplane, blades for jet engines or power plant turbines, and other parts that have continuously varying shapes.

Removed sections can also be drawn to a larger scale for better representation of the details of the cross section and for dimensioning. The scale used for the removed section view is labeled beneath the view. Sometimes removed sections are placed on center lines adjacent to the axis of revolution.

3.4.6 Offset Sections

An **offset section** has a cutting plane that is bent at one or more 90-degree angles to pass through important features.

(A) Revolved section

(B) Revolved section; broken view

Figure 3.14   Revolved Section
A revolved section view is created by passing a cutting plane through the object, then revolving the cross section 90 degrees.

(Figure 3.16A) Offset sections are used for complex parts that have a number of important features that cannot be sectioned using a straight cutting plane. In Figure 3.16, the cutting plane is first bent at 90 degrees to pass through the hole and then bent at 90 degrees to pass through the slot. The front portion of the object is "removed" to create a

Figure 3.15   Removed Section
A removed section view is created by making a cross section, then moving it to an area adjacent to the view.

full section view of the part. The cutting plane line is drawn with 90-degree offsets, as shown in Figure 3.16A. As shown in Figure 3.16B, the change of plane that occurs when the cutting plane is bent at 90 degrees is not represented with lines in the section view.

3.4.7 Assembly Sections

Assembly sections are typically orthographic, pictorial, full or half section views of parts as assembled. Leonardo da Vinci was one of the first to create assembly sections, using them to illustrate pictorially the complex machine designs he developed.

Section assembly drawings follow special conventions. Standard parts, such as fasteners, dowels, pins, washers, springs, bearings, and gears, and nonstandard parts, such as shafts, are *not* sectioned; instead, they are drawn showing all of their exterior features. For example, in Figure 3.17, fasteners would be cut in half by the cutting plane, yet they are not cross-hatched with section lines.

Typically, the following features are not section lined in a mechanical assembly section drawing:

Shafts.	Ribs.
Bearings, roller or ball.	Spokes.
Gear teeth.	Lugs.
Threaded fasteners.	Washers.
Nuts and bolts.	Keys.
Rivets.	Pins.

Adjacent parts in assembly sections are cross-hatched at different angles so that they are more easily identified. (Figure 3.18) Different material symbols can also be used for this purpose. Also, if a part in an assembly section is separated by some distance, the section lines are still drawn in the same direction.

CAD 3-D modeling software can create models of each part, the individual models can be sectioned, and the section views can be placed together, resulting in a 3-D assembly section. An alternative involves using a feature that adds translucency to parts to reveal interior assemblies. (Figure 3.19) The models can be used to check for interference of parts or can be analyzed by dynamically showing the parts in action. The models can also be rotated to produce an orthographic assembly section view.

Normal multiview

(A) Offset section view

(B) No!

DO NOT SHOW!

Figure 3.16   Offset Section

An offset section view is created by bending the cutting plane at 90-degree angles to pass through important features.

Figure 3.17   Standard Parts Not Section Lined

Standard parts, such as fasteners and shafts, are not section lined in assembly sections, even if they are cut by the cutting plane.

Adjacent parts

Different material symbol

Same part

Figure 3.18   Section Lining Adjacent Parts

Adjacent parts in an assembly section are section lined at different angles so that individual parts can be more easily identified.

Figure 3.19 Translucency of a CAD Model
With a 3-D CAD model, translucency can be used instead of cutting planes to reveal interior features. Courtesy of Hewlett-Packard.

3.5 | AUXILIARY VIEW PROJECTION THEORY

An **auxiliary view** is an orthographic view that is projected onto any plane other than one of the six principal views. Figure 3.20A shows three principal views of an object. Surface ABCD is in an inclined plane and is therefore never seen in true size or shape in any of these views. In a multiview drawing, a true size and shape plane is shown only when the *line of sight* (LOS) used to create the view

is perpendicular to the projection plane. To show the true size and shape of surface ABCD, an auxiliary view can be created by positioning a line of sight perpendicular to the inclined plane, then constructing the new view. (Figure 3.20B) Two methods of creating auxiliary views are the *fold-line method* and the *reference plane method*. These are discussed in the following sections. The use of construction techniques for auxiliary views has diminished with the use of 3-D CAD.

3.5.1 Fold-Line Method

In Figure 3.21, the object is suspended in a glass box to show the six principal views, created by projecting the object onto the planes of the box. The box is then unfolded, resulting in the six principal views. However, when the six views are created, surface ABCD never appears true size and shape; it always appears either foreshortened or on edge.

Figure 3.22 shows the object suspended inside a glass box, which has a special or *auxiliary plane* that is parallel to inclined surface ABCD. The line of sight required to create the auxiliary view is perpendicular to the new projection plane and to surface ABCD. The auxiliary plane is perpendicular to and hinged to the frontal plane, creating a *fold line* between the front view and the new auxiliary view.

In Figure 3.23, the auxiliary glass box is unfolded, with the fold lines between the views shown as phantom lines. In the auxiliary view, surface ABCD is shown true size and shape and is located at distance M from the fold line.

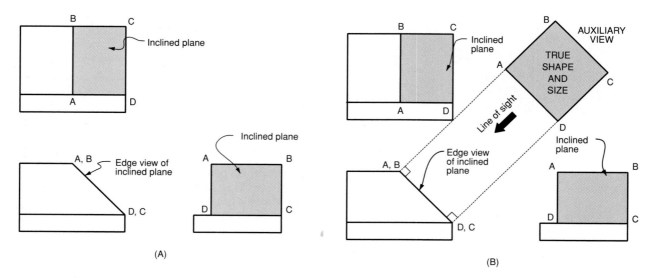

(A)

(B)

Figure 3.20 Auxiliary View
An auxiliary view of an inclined plane is not one of the principal views.

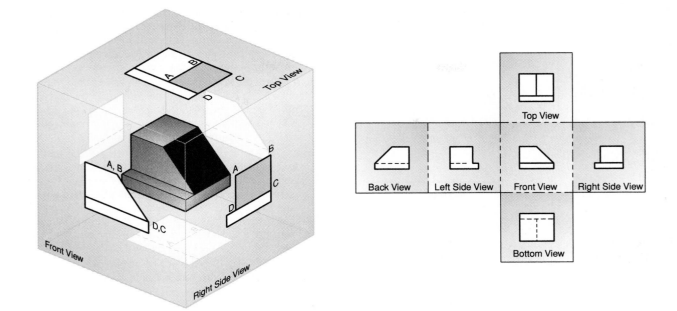

Figure 3.21   Object in Glass Box, and Resulting Six Views When the Box Is Unfolded

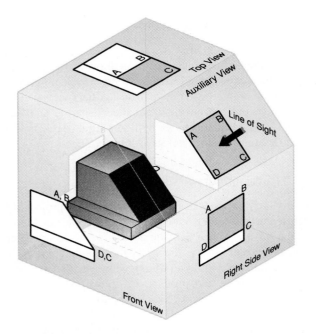

Figure 3.22   Object in Glass Box with Special Auxiliary Plane

Figure 3.23   Unfolding the Glass Box to Create an Auxiliary View of the Inclined Plane

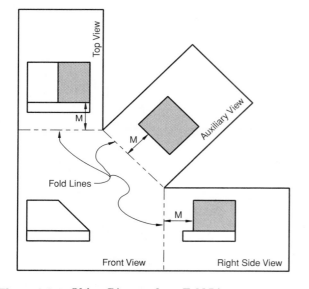

Figure 3.24   Object Distance from Fold Line
Object distance from the frontal plane determines the distance from the fold lines in the right side, auxiliary, and top views.

The line AB in the top view is also located at distance M from its fold line. Changing the position of the object, such as moving it closer to the frontal plane, changes distance M. (Figure 3.24)

3.5.2 Constructing an Auxiliary View

Figure 3.25 shows an auxiliary view that is projected from the front view of an object, using the fold-line method. Since plane ABCD is an inclined plane in the principal views, an auxiliary view is needed to create a true-size view of that plane. An auxiliary view of plane ABCD is created as described in the following steps.

Constructing an Auxiliary View

Step 1.   Given the front, top, and right side views, draw fold line F–1 using a phantom line parallel to the edge view of the inclined surface. Place line F–1 at any convenient distance from the front view.

Step 2.   Draw fold line F–H between the front and top views. Line F–H should be perpendicular to the projectors between the views and at a distance X from the rear edge of the top view. Draw fold line F–P between the front and right side views, perpendicular to the projectors between

the two views and at the distance X from the rear edge of the right side view. The distance from fold line F–H to the top view must be equal to the distance from fold line F–P to the right side view. Draw parallel projectors between the principal views, using construction lines.

Step 3.   Project the length of the inclined surface from the front view to the auxiliary view, using construction lines. The projectors are *perpendicular* to the edge view and projected well into the auxiliary view from corners A,B and D,C.

Step 4.   Transfer the depth of the inclined surface from the top view to the auxiliary view by first measuring the *perpendicular* distance from fold line H–F to point C at the rear of the top view. This is distance X. Measure this same distance on the projectors in the auxiliary view, measuring from fold line F–1. The measurement used to locate point C could have been taken from the profile view.

Step 5.   From point C in the auxiliary view, draw a line perpendicular to the projectors. Depth dimension Y is transferred from the top view by measuring the *perpendicular* distance from fold line H–F to point A (or D) in the top view and transferring that distance to the auxiliary view along the projectors *perpendicular* to fold line F–1. Draw a line at the transferred point A (or D) in the auxiliary view, perpendicular to the projectors.

Step 6.   To complete the auxiliary view of the inclined surface, darken lines AB and DC.

Only the inclined plane has been drawn in the auxiliary view; the rest of the object is not represented. When only the feature of interest is drawn in an auxiliary view and not the whole object, the view is called a *partial auxiliary view.* Most auxiliary views will be partial. Also, hidden features are not shown unless absolutely necessary.

3.5.3 Partial Auxiliary Views

In auxiliary views, it is normal practice *not* to project hidden features or other features that are not part of the inclined surface. When only the details for the inclined surface are projected and drawn in the auxiliary view, the view is called a **partial auxiliary view.** A partial auxiliary view saves time and produces a drawing that is much more readable. Figure 3.26 shows a partial and a full auxiliary view of the same object. The full auxiliary view is harder to draw, read, and visualize. In this example, some of the holes have to be drawn as ellipses in the full auxiliary view. Sometimes a break line is used in a partial auxiliary view. When drawing break lines, do not locate them coincident with a visible or hidden line.

Figure 3.25 **Constructing a Depth Auxiliary View to Determine the True Size and Shape of the Inclined Surface**

3.5.4 Half Auxiliary Views

Symmetrical objects can be represented as a **half auxil-iary view;** that is, only half of the object is drawn. The construction of a half auxiliary view is the same as described earlier for full auxiliary views. Figure 3.27 shows an object represented as a half auxiliary view.

Figure 3.26   A Full Auxiliary View, Including Hidden Lines, and a Partial Auxiliary View with No Hidden Lines

Figure 3.27   A Half Auxiliary View of a Symmetrical Feature

Questions for Review

1. Define section drawings.
2. Describe how 3-D CAD is used to create section views.
3. Describe how cutting plane lines are used.
4. Sketch the two standard types of cutting plane lines.
5. List three applications of section views.
6. What convention is used for hidden lines in a section view?
7. What convention is used for hidden lines on the unsectioned half of a half section view?
8. Define section lines.
9. Sketch the material symbol used to represent steel.
10. Describe the difference between a revolved and a removed section view.
11. List some of the standard parts that are not sectioned in an assembly section. Explain why.
12. What type of line is used to separate the sectioned half from the unsectioned half of a half section view?
13. Define auxiliary views.
14. What is a partial auxiliary view?

Problems

3.1 For Figure 3.28 section view problems, sketch (or draw with instruments or use a CAD software package) the orthographic view containing the cutting plane line and appropriate section view. Assume each grid represents 6 mm.

3.2 Sketch, or draw with instruments or CAD, the necessary views, including a section view, of the objects shown in Figures 3.29 through 3.36.

3.3 Using instruments of CAD, sketch or draw the two given views and a partial auxiliary view with hidden lines of the largest inclined surface in Figure 3.37.

3.4 Using instruments of CAD, sketch or draw the two given views and a complete auxiliary view without hidden lines of the largest inclined surface in Figure 3.38 objects.

3.5 For Figures 3.39 – 3.42, sketch (or draw with instruments or use a CAD software package) two adjacent orthographic views of the object and a projected auxiliary view showing the object such that the largest inclined or largest oblique plane is in true shape.

Figure 3.28   Section View Problems

Figure 3.28 Continued

Figure 3.29   Ring Collar

FILLETS & ROUNDS
R .62 U.O.S.

Figure 3.30   Bracket

FILLETS & ROUNDS R .4

Figure 3.31   Counter Block

Figure 3.32 Center Support

Figure 3.33 Taper Collar

FILLETS & ROUNDS R .25
U.O.S.

Figure 3.34 Pulley

Figure 3.35   Shoulder Stop

FILLETS & ROUND R .13

Figure 3.36   Mounting

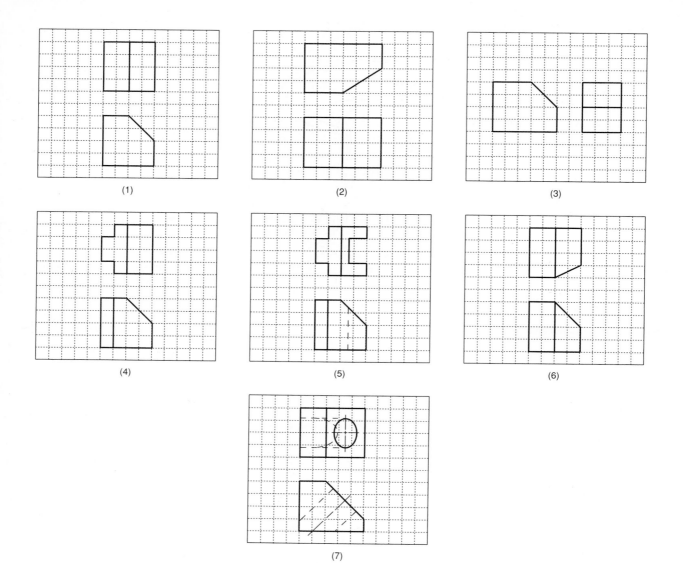

Figure 3.37   Objects with Inclined Surfaces for Problem 3.3

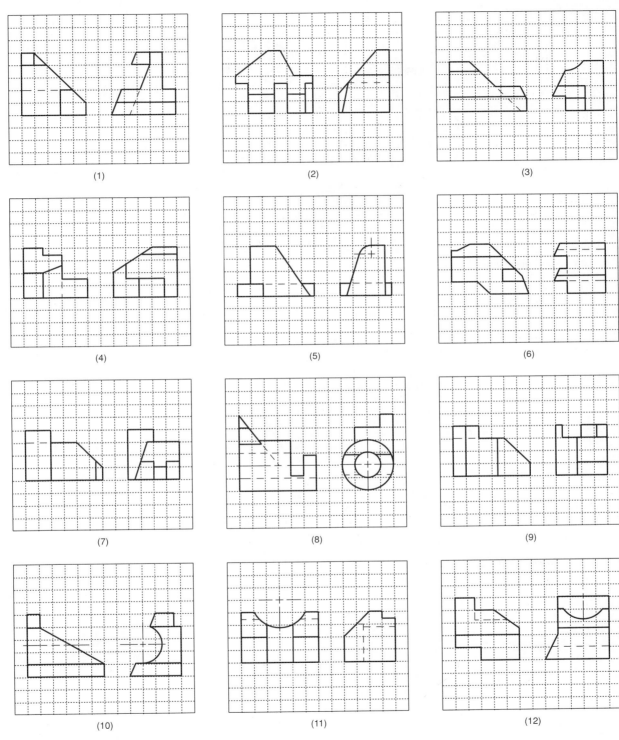

(1)

(2)

(3)

(4)

(5)

(6)

(7)

(8)

(9)

(10)

(11)

(12)

Figure 3.38   **Multiviews for Problem 3.4**

Figure 3.39   Automatic Stop

METRIC

Figure 3.40   Slotted Guide

ALL ANGLES 90 DEGREES UNLESS
OTHERWISE NOTED

Figure 3.41   Oblique Support

Figure 3.42   Harness Base

Problem 3.1

Refer to Figure 3.28. Sketch the necessary section views. Divide the long dimension of this problem sheet with a thick dark line. A section view drawing can fit on either half of the paper.

Sketch Number: _____

Name:_____

Div/Sec: _____

Date:_____

Problem 3.1

Refer to Figure 3.28. Sketch the necessary section views. Divide the long dimension of this problem sheet with a thick dark line. A section view drawing can fit on either half of the paper.

Sketch Number: _____

Name:_____

Div/Sec: _____

Date:_____

Problem 3.1

Refer to Figure 3.28. Sketch the necessary section views. Divide the long dimension of this problem sheet with a thick dark line. A section view drawing can fit on either half of the paper.

Sketch Number: _____

Name:_____

Div/Sec: _____

Date:_____

Problem 3.2

Refer to Figures 3.29–3.36. Sketch or draw the necessary section views.

Sketch Number: _____

Name: _____

Div/Sec: _____

Date: _____

Problem 3.2

Refer to Figures 3.29–3.36. Sketch or draw the necessary section views.

Sketch Number: _____

Name: _____

Div/Sec: _____

Date: _____

Problem 3.2

Refer to Figures 3.29–3.36. Sketch or draw the necessary section views.

Sketch Number: _____

Name:_____

Div/Sec: _____

Date:_____

Problem 3.3

Refer to Figure 3.37. Sketch or draw the two given views and a partial auxiliary view of the inclined surface.

Sketch Number: _____

Name:_____

Div/Sec: _____

Date:_____

Problem 3.3

Refer to Figure 3.37. Sketch or draw the two given views and a partial auxiliary view of the inclined surface.

Sketch Number: _____

Name:_____

Div/Sec: _____

Date:_____

Problem 3.4

Refer to Figure 3.38. Sketch or draw the two given views and a partial auxiliary view of the inclined surface.

Sketch Number: _____

Name: _____

Div/Sec: _____

Date: _____

Problem 3.4

Refer to Figure 3.38. Sketch or draw the two given views and a partial auxiliary view of the inclined surface.

Sketch Number: _____

Name:_____

Div/Sec: _____

Date:_____

Problem 3.4

Refer to Figure 3.38. Sketch or draw the two given views and a partial auxiliary view of the inclined surface.

Sketch Number: _____

Name:_____

Div/Sec: _____

Date:_____

Problem 3.5

Refer to Figures 3.39 through 3.42. Sketch or draw the two given views and a partial auxiliary view of the inclined surface.

Sketch Number: _____

Name:_____

Div/Sec: _____

Date:_____

Problem 3.5

Refer to Figures 3.39 through 3.42. Sketch or draw the two given views and a partial auxiliary view of the inclined surface.

Sketch Number: _____

Name: _____

Div/Sec: _____

Date: _____

Chapter

Dimensioning and Tolerancing Practices

4

OBJECTIVES

After completing this chapter, you will be able to:

1. Apply the standard dimensioning practices for mechanical drawings.
2. Differentiate between current ASME standards and past practices for dimensioning.
3. Apply English and metric tolerances to dimensions.
4. Calculate standard tolerances for precision fits.
5. Apply tolerances using the basic shaft and basic hole systems.

4.1 | DIMENSIONING

Geometrics is the science of specifying and tolerancing the shapes and locations of features on objects. In design work, it is essential that the roundness of a shaft be as clearly stated as the size. Once the shape of a part is defined with an orthographic drawing, the size information is added in the form of **dimensions.** Dimensioning a drawing also identifies the tolerance (or accuracy) required for each dimension. If a part is dimensioned properly, then the intent of the designer is clear to both the person making the part and the inspector checking the part. Everyone in this *circle of information* (design, manufacturing, quality control) must be able to speak and understand a common language. A well-dimensioned part is a component of this communications process.

4.2 | SIZE AND LOCATION DIMENSIONS

A well-dimensioned part or structure will *communicate* the size and location requirements for each feature. *Communications* is the fundamental purpose of dimensions.

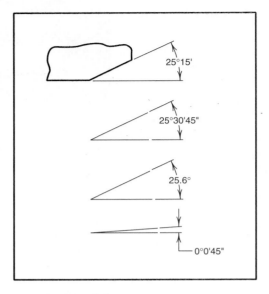

Figure 4.1   Angular Units
Angular dimensions are shown either in decimal degrees or in degrees, minutes, and seconds.

Designs are dimensioned based on two criteria:

1. Basic sizes and locations of features.
2. Details for construction and for manufacturing.

4.2.1 Units of Measure

The unit of measurement selected should be in accordance with the policy of the user. Construction and architecture drawings use feet and inches for dimensioning units. On a drawing for use in American industry for manufacturing, all dimensions are in inches, unless otherwise stated. Generally, if parts are more than 72 inches in length, the drawings will switch to feet and inches for the standard unit of measure. Most countries outside of the United States use the metric system of measure, or the international system of units (SI), which is based on the meter.

The SI system is being used more in the United States because of global trade and multinational company affiliations. The common metric unit of measure on engineering drawings is the *millimeter,* abbreviated as *mm.*

Angular dimensions are shown either in decimal degrees or in degrees, minutes, and seconds. The symbol used for degrees is °, for minutes ', and for seconds ". Where only minutes and seconds are specified, the number of minutes or seconds are preceded by the 0°. Figure 4.1 shows examples of angular units used to dimension angles.

4.2.2 Terminology

There are a number of terms important to dimensioning practices. These terms are illustrated in Figure 4.2 and are defined as follows:

1. **Dimension**—the numerical value that defines the size, shape, location, surface texture, or geometric characteristic of a feature. Normally, dimension text is 3 mm or 0.125″ high, and the space between lines of text is 1.5 mm or 0.0625″. (Figure 4.3)

2. **Basic dimension**—a numerical value defining the theoretically exact size, location, profile, or orientation of a feature relative to a coordinate system established by datums. They are identfied on a drawing by enclosing the dimension in a rectangular box. Basic dimensions have no tolerance. They locate the perfect geometry of a part, while the acceptable variation or geometric tolerance is described in a feature control frame.

3. **Reference dimension**—a numerical value enclosed in parentheses, providing for information only and not directly used in the fabrication of the part. A reference dimension is a calculated size without a tolerance used to show the intended design size of a part.

4. **Dimension line**—a thin, solid line that shows the extent and direction of a dimension. Dimension lines are broken for insertion of the dimension numbers.

5. **Arrows**—symbols placed at the ends of dimension lines to show the limits of the dimension, leaders, and cutting plane lines. Arrows are uniform in size and style, regardless of the size of the drawing. Arrows are usually about 3 mm ($\frac{1}{8}$″) long and should be one-third as wide as they are long. (Figure 4.4)

6. **Extension line**—a thin, solid line perpendicular to a dimension line, indicating which feature is associated with the dimension.

7. **Visible gap**—there should be a visible gap of 1 mm ($\frac{1}{16}$″) between the feature's corners and the end of the extension line.

8. **Leader line**—a thin, solid line used to indicate the feature with which a dimension, note, or symbol is associated. Leader lines are generally a straight line drawn at an angle that is neither horizontal nor vertical. Leader lines are terminated with an arrow touching the part or detail. On the end opposite the arrow, the leader line will have a short, horizontal shoulder (3 mm or .125″ long). Text is extended from this shoulder such that the text height is centered with the shoulder line. Two or more adjacent leaders on a drawing should be drawn parallel to each other.

Figure 4.2   Important Elements of a Dimensioned Drawing

Decimal dimensioning

Millimeter dimensioning

Figure 4.3   Text Height and Standards Used for Decimal and Millimeter Dimensioning

3mm or .125"

W

3W

Filled   Closed   Open

Figure 4.4   Dimensions Used to Draw an Arrowhead
Arrows are three times as long as they are wide.

9. **Limits of size**—the largest acceptable size and the minimum acceptable size of a feature. The value for the largest acceptable size, expressed as the maximum material condition (MMC), is placed over the value for the minimum acceptable size, expressed as the least material condition (LMC), to denote the limit-dimension-based tolerance for the feature.

10. **Plus and minus dimension**—the allowable positive and negative variance from the dimension specified. The plus and minus values may or may not be equal.

11. **Diameter symbol**—a symbol that precedes a numerical value, to indicate that the dimension shows

Figure 4.5
Size and location dimensions are used to describe parts for manufacture.

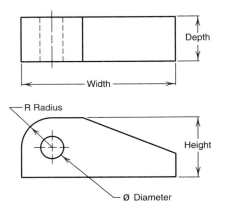

Figure 4.6 **Dimensions Showing the Sizes of Features, Such as the Height and Depth of the Part and the Diameter of the Hole**

the diameter of a circle. The symbol used is the Greek letter phi (ø).

12. **Radius symbol**—a symbol that precedes a numerical value to indicate that the associated dimension shows the radius of a circle. The radius symbol used is the capital letter R.

13. **Tolerance**—the amount that a particular dimension is allowed to vary. All dimensions (except reference dimensions) have an associated tolerance. A tolerance may be expressed either through limit dimensioning, plus and minus dimensioning, or a general note. The tolerance is the difference between the maximum and minimum limits.

4.2.3 Basic Concepts

A size dimension might be the overall width of a part or structure, or the diameter of a drilled hole. (Figure 4.5) A location dimension might be the length from the edge of an object to the center of a feature. The basic criterion is, "What information is necessary to manufacture or construct the object?" For example, to drill a hole, the manufacturer would need to know the diameter of the hole, the location of the center of the hole, and the depth to which the hole is to be drilled. These three dimensions describe the hole in sufficient detail for the feature to be made using machine tools.

4.2.4 Size Dimensions

Horizontal—the left-to-right distance relative to the drawing sheet. In Figure 4.6, the *width* is the only horizontal size dimension.

Vertical—the up and down distance relative to the drawing sheet. In Figure 4.6, the *height* and the *depth* are both vertical dimensions, even though they are in two different directions on the part.

Diameter—the full distance across a circle, measured through the center. This dimension is usually used only on full circles or on arcs that are more than half of a full circle.

Radius—the distance from the center of an arc to any point on the arc, usually used on arcs less than half circles. In Figure 4.6, the radius points to the outside of the arc, even though the distance measured is to the center, which is inside. The endpoints of the arc are tangent to the horizontal and vertical lines, making a quarter of a circle. This is assumed, and there is no need to note it. If the radius is not positioned in this manner, then the actual center of the radius must be located.

4.2.5 Location and Orientation Dimensions

- *Horizontal position*—In Figure 4.7, dimensions A and D are horizontal position dimensions that locate the beginnings of the angle. Dimension A measures more than one feature—the sum of the arc's radius and the straight line. The measurement for dimension A is taken parallel to the dimension line. Dimension D is the measurement of a single feature—the sloping line—but it is not the true length of the line. Rather, it is the left-to-right distance that the line displaces. This is called the "delta X

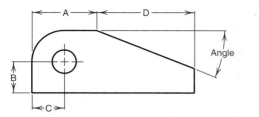

Figure 4.7 Dimensions Showing the Location and Orientations of Features, Such as the Location of the Center of the Hole

Decimal dimensioning

Millimeter dimensioning

Figure 4.8 Dimension Text Placement
Standard practice for the placement of dimensions depends on the space available.

value," or the change in the X direction. The C dimension measures the horizontal location of the center of the hole and arc.

- *Vertical position*—The B dimension in Figure 4.7 measures the vertical position of the center of the hole. For locating the hole, the dimensions are given to the center, rather than the edges of the hole. All circular features are located from their centers.
- *Angle*—The angle dimension in Figure 4.7 gives the angle between the horizontal plane and the sloping surface. The angle dimension can be taken from several other directions, measuring from any measurable surface.

4.2.6 Standard Practices

The guiding principle for dimensioning a drawing is *clarity*. To promote clarity, ANSI developed standard practices for showing dimensions on drawings.

Placement Dimension placement depends on the space available between extension lines. When space permits, dimensions and arrows are placed *between* the extension lines, as shown in Figures 4.8A and E.

When there is room for the numerical value but not the arrows as well, the value is placed between the extension lines and the arrows are placed outside the extension lines, as shown in Figures 4.8B and F.

When there is room for the arrows but not the numerical value, the arrows are placed between the extension lines, and the value is placed outside the extension lines and adjacent to a leader, as shown in Figures 4.8C and G.

When the space is too small for either the arrows or the numerical value, both are placed outside the extension lines, as shown in Figures 4.8D and H.

Spacing The minimum distance from the object to the first dimension is 10 mm (⅜″), as shown in Figure 4.9.

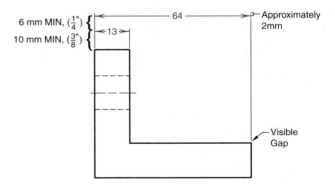

Figure 4.9 Minimum Dimension Line Spacing
Standard practice for the spacing of dimensions is 10 mm from the view and 6 mm between dimension lines.

The minimum spacing between dimensions is 6 mm (¼″). These values may be increased where appropriate.

There should be a visible gap between an extension line and the feature to which it refers. Extension lines should extend about 2–3 mm (⅛″) beyond the last dimension line.

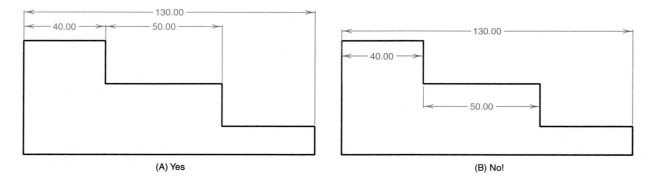

(A) Yes   (B) No!

Figure 4.10 **Group Dimensions**
In standard practice, dimensions are grouped on a drawing. Do not use object lines as extension lines for a dimension.

Figure 4.11 **Stagger Dimension Text**
The general practice is to stagger the dimension text on parallel dimensions.

Figure 4.12 **Extension Line Practice**
Extension lines should not cross dimension lines, are not broken when crossing object or other extension lines, and are broken when crossing arrows.

Grouping and Staggering Dimensions should be grouped for uniform appearance, as shown in Figure 4.10. As a general rule, do not use object lines as part of your dimension. (Figure 4.10B) Where there are several parallel dimensions, the values should be staggered, as shown in Figure 4.11.

Extension Lines **Extension lines** are used to relate a dimension to one or more features and are usually drawn

perpendicular to the associated dimension line. Where angled extension lines are used, they must be parallel, and the associated dimension lines must be drawn in the direction to which they apply.

Extension lines should not cross dimension lines, and should avoid crossing other extension lines whenever possible. When extension lines cross object lines or other extension lines, they should not be broken. When extension lines cross or are close to arrowheads, they should be broken for the arrowhead. (Figure 4.12)

When the center of a feature is being dimensioned, the centerline of the feature is used as an extension line. (Figure 4.13A) When a point is being located by extension lines only, the extension lines must pass through the point. (Figure 4.13B)

Centerline used as an extension line

20

25

(A)

48

12

41

(B)

Figure 4.13
The center of a feature, such as a hole, is located by making the center lines extension lines for the dimension. Extension lines can also cross to mark a theoretical point.

2.00
1.00
2.00
1.00

Unidirectional
Current standard

2.00
1.00
2.00
1.00

Aligned
Old standard

Figure 4.14   Unidirectional and Aligned Methods
The unidirectional method of placing text is the current standard practice.

(A) No!

(B) Yes

Figure 4.15   Dimension Outside of View
Dimensions are kept off the view, unless necessary for clarity.

Reading Direction   All dimensions and note text must be oriented to read from the bottom of the drawing (relative to the drawing format). This is called **unidirectional dimensioning.** (Figure 4.14) The *aligned method* of dimensioning may be seen on older drawings or architectural drawings but is not approved by the current ANSI standards. **Aligned dimensions** have text placed parallel to the dimension line, with vertical dimensions read from the right of the drawing sheet.

View Dimensioning   Dimensions are to be kept outside the boundaries of views, wherever practical. (Figure 4.15B) Dimensions may be placed within the boundaries where extension or leader lines would be too long or where clarity would be improved.

Repetitive Features   The symbol $\times$ is used to indicate the number of times a feature is to be repeated. The number of repetitions, followed by the symbol $\times$ and a space, pre-

4X Ø .375

Figure 4.16   Using the Symbol $\times$ to Dimension Repetitive Features

cedes the dimension text. For example, in Figure 4.16, 4 $\times$ ø .375 means that there are 4 holes with a diameter of .375″.

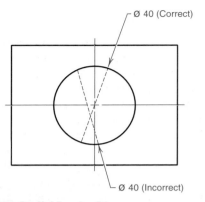

Figure 4.17 Radial Leader Lines

Leader lines used to dimension holes must be radial.

Figure 4.18 Symbols for Drilling Operations

Symbols to represent machine drilling operations always precede the diameter symbol.

4.3 | DETAIL DIMENSIONING

Holes are typically dimensioned in a view that best describes the shape of the hole. For diameters, the diameter symbol must precede the numerical value. When holes are dimensioned with a leader line, the leader line must be radial. (Figure 4.17) A **radial line** is one that passes through the center of a circle or arc if extended. If it is not clear whether a hole extends completely through a part, the word THRU can follow the numerical value.

Symbols are used for spotface, counterbored, and countersunk holes. These symbols must always precede the diameter symbol. (Figure 4.18) The depth symbol is used to indicate the depth of a hole. The depth symbol precedes the numerical value. When the depth of a blind hole is specified, the depth is to the full diameter of the hole and not to the point. (Figure 4.19) When a chamfer or countersink is placed in a curved surface, the diameter refers to the minimum diameter of the chamfer or countersink.

Slotted holes may be dimensioned in any of several ways, depending on which is most appropriate for the application. The various options for dimensioning slotted holes are shown in Figure 4.20.

Figure 4.19 Dimensioning a Blind Hole

The depth of a blind hole reflects the depth of the full diameter.

Figure 4.20 Dimensioning Slots
Several methods are appropriate for dimensioning slots.

4.3.1 Diameter versus Radius

If a full circle or an arc of more than half a circle is being dimensioned, the diameter is specified, preceded by the diameter symbol, which is the Greek letter phi. If the arc is less than half a circle, the radius is specified, preceded by an R. Concentric circles are dimensioned in the longitudinal view, whenever practical. (Figure 4.21)

As previously stated, radii are dimensioned with the radius symbol preceding the numerical value. The dimension line for radii shall have a single arrowhead touching the arc. When there is adequate room, the dimension is placed between the center of the radius and the arrowhead. (Figure 4.22) When space is limited, a radial leader line is used. When an arc is not clearly defined by being tangent to other dimensioned features on the object, the center of the arc is noted with a small cross. (Figure 4.22) The center is not shown if the arc is tangent to other defined surfaces.

4.3.2 Dimensioning Guidelines

The importance of accurate, unambiguous dimensioning cannot be overemphasized. There are many cases where

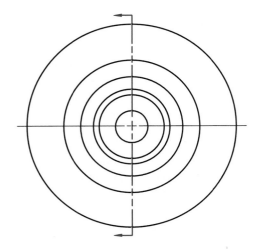

Figure 4.21 Dimensioning Concentric Circles
Concentric circles are dimensioned in the longitudinal view.

Figure 4.22 Dimensioning Arcs
Arcs of less than half a circle are dimensioned as radii, with the R symbol preceding the dimension value.

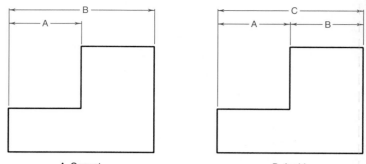

A. Correct   B. Avoid

Figure 4.23 Avoid Overdimensioning
Double dimensioning can cause problems because of tolerancing.

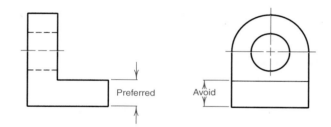

Figure 4.24 Dimension the Most Descriptive View
Dimensions are placed in the most descriptive or contour view.

an incorrect or unclear dimension has added considerable expense to the fabrication of a product, caused premature failure, or, in some cases, caused loss of life. *The primary guideline is clarity: Whenever two guidelines appear to conflict, the method that most clearly communicates the size information shall prevail.* Use the following dimensioning guidelines:

1. Every dimension must have an associated tolerance, and that tolerance must be clearly shown on the drawing.

2. Double dimensioning of a feature is not permitted. For example, in Figure 4.23, there are two ways to arrive at the overall length of the object: by adding dimensions A and B or by directly measuring the dimension C. Since each dimension must have a tolerance, it is not clear which tolerance would apply to the overall length: the tolerance for dimension C or the sum of the tolerances for dimensions A and B. This ambiguity can be eliminated by removing one of the three dimensions. The dimension that has the least importance to the function of the part should be left out. In this case, dimension A would probably be deleted.

3. Dimensions should be placed in the view that most clearly describes the feature being dimensioned (contour dimensioning). For example, Figure 4.24 illustrates a situation in which the height of a step is being dimensioned. In this case, the front view more clearly describes the step feature.

4. Maintain a minimum spacing between the object and the dimension and between multiple dimensions. This spacing is shown in Figure 4.9. If the spacing is reduced, the drawing will be more difficult to read and a lack of clarity will result.

5. A visible gap shall be placed between the ends of extension lines and the feature to which they refer.

6. Manufacturing methods should not be specified as part of the dimension, unless no other method of manufacturing is acceptable. The old practice of using *drill* or *bore* is discouraged. Because a drawing becomes a legal document for manufacturing, specifying inappropriate manufacturing methods can cause unnecessary expense and may trigger litigation.

7. Avoid placing dimensions within the boundaries of a view, whenever practicable. If several dimensions are placed in a view, differentiation between

Table 4.1 Basics of Dimensioning

Parts of a Dimension

Dimension—A dimension is a numerical value shown on a drawing to define the size of an object or a part of an object. Dimensions may be expressed in either U.S. or metric units.

Dimension line—A dimension line is a thin solid line used to show the extent and direction of a dimension.

Arrowheads—Arrowheads are placed at the ends of dimension lines to show the limits of the dimension.

Extension line—Extension lines are thin lines drawn perpendicular to dimension lines, and they indicate the feature of the object to which the dimension refers.

Leader line—A leader line is a thin solid line used to direct dimensions or notes to the appropriate feature.

Tolerance—Tolerances are the amount a dimension is allowed to vary. The tolerance is the difference between the maximum and minimum permitted sizes.

Principles of Good Dimensioning

The overriding principle of dimensioning is clarity.

1. Each feature of an object is dimensioned once and only once.
2. Dimensions should be selected to suit the function of the object.
3. Dimensions should be placed in the most descriptive view of the feature being dimensioned.
4. Dimensions should specify only the size of a feature. The manufacturing method should only be specified if it is a mandatory design requirement.
5. Angles shown on drawings as right angles are assumed to be 90 degrees unless otherwise specified, and they need not be dimensioned.
6. Dimensions should be located outside the boundaries of the object whenever possible.
7. Dimension lines should be aligned and grouped where possible to promote clarity and uniform appearance.
8. Crossed dimension lines should be avoided whenever possible. When dimension lines must cross, they should be unbroken.
9. The space between the first dimension line and the object should be at least ⅜ inch (10 mm).
   The space between dimension lines should be at least ¼ inch (6 mm).
10. There should be a visible gap between the object and the origin of an extension line.
11. Extension lines should extend ⅛ inch (3 mm) beyond the last dimension line.
12. Extension lines should be broken if they cross or are close to arrowheads.
13. Leader lines used to dimension circles or arcs should be radial.
14. Dimensions should be oriented to be read from the bottom of the drawing.
15. Diameters are dimensioned with a numerical value preceded by the diameter symbol.
16. Concentric circles should be dimensioned in a longitudinal view whenever possible.
17. Radii are dimensioned with a numerical value preceded by the radius symbol.
18. When a dimension is given to the center of an arc or radius, a small cross is shown at the center.
19. The depth of a blind hole may be specified in a note. The depth is measured from the surface of the object to the deepest point where the hole still measures a full diameter in width.
20. Counterbored, spotfaced, or countersunk holes should be specified in a note.

the object and the dimensions may become difficult.

8. Dimensions for materials typically manufactured to gages or code numbers shall be specified by numerical values. The gages or code numbers may be shown in parentheses following the numerical values.

9. Unless otherwise specified, angles shown in drawings are assumed to be 90 degrees.

10. Avoid dimensioning hidden lines. Hidden lines are less clear than visible lines.

11. The depth and diameter of blind, counterbored, or countersunk holes may be specified in a note. (Figures 4.18 and 4.19)

12. Diameters, radii, squares, counterbores, spotfaces, countersinks, and depths should be specified with the appropriate symbol preceding the numerical value. (Figure 4.18)

13. Leader lines for diameters and radii should be radial lines. (Figure 4.17)

Table 4.1 summarizes the basics of dimensioning.

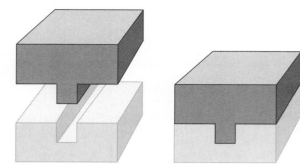

Figure 4.25

A system is two or more mating parts.

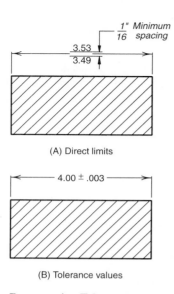

(A) Direct limits

(B) Tolerance values

Figure 4.26   Representing Tolerance Values

Tolerances are represented as direct limits or as tolerance values.

4.4 | TOLERANCING

Tolerances are used to control the variation that exists on all manufactured parts. Toleranced dimensions control the amount of variation on each part of an assembly. The amount each part is allowed to vary depends on the function of the part and of the assembly. For example, the tolerances placed on electric hand-drill parts are not as stringent as those placed on jet engine parts. The more accuracy needed in the machined part, the higher the manufacturing cost. Therefore, tolerances must be specified in such a way that a product functions as it should at a cost that is reasonable.

A tolerance of 4.650 ± .003 means that the final measurement of the machined part can be anywhere from 4.653 to 4.647 and the part would still be acceptable. The lower and upper allowable sizes are referred to as the *limit dimensions,* and the *tolerance* is the difference between the limits. In the example, the **upper limit** (largest value) for the part is 4.653, the **lower limit** (smallest value) is 4.647, and the tolerance is .006.

Tolerances are assigned to mating parts. For example, the slot in the part shown in Figure 4.25 must accommodate another part. A *system* is two or more mating parts.

4.5 | TOLERANCE REPRESENTATION

Tolerance is the total amount a dimension may vary and is the difference between the maximum and minimum limits. Because it is impossible to make everything to an exact size, tolerances are used on production drawings to control the manufacturing process more accurately and to

Figure 4.27   Geometric Tolerance System Used to Dimension Parts

control the variation between mating parts. ASME standard Y14.5M-1994 is commonly used in U.S. industries to specify tolerances on engineering drawings.

Tolerances can be expressed in several ways:

1. Direct limits, or as tolerance values applied directly to a dimension. (Figure 4.26)
2. Geometric tolerances. (Figure 4.27)
3. Notes referring to specific conditions.
4. A general tolerance note in the title block.

4.5.1 General Tolerances

General tolerances are given in a note or as part of the title block. A general tolerance note would be similar to:

ALL DECIMAL DIMENSIONS TO BE HELD
TO ± .002″

This means that a dimension such as .500 would be assigned a tolerance of ± .002, resulting in an upper limit of .502 and a lower limit of .498.

4.5.2 Limit Dimensions

Tolerances can be applied directly to dimensioned features, using limit dimensioning. This is the ASME preferred method; the maximum and minimum sizes are specified as part of the dimension. (See Figure 4.26A.) Either the upper limit is placed above the lower limit, or, when the dimension is written in a single line, the lower limit precedes the upper limit, and they are separated by a dash. A minimum spacing of 1/16 inch is required when the upper limit is placed above the lower limit.

4.5.3 Plus and Minus Dimensions

With this approach, the basic size is given, followed by a plus/minus sign and the tolerance value. (Figure 4.28) Tolerances can be unilateral or bilateral. A **unilateral tolerance** varies in only one direction. A **bilateral tolerance** varies in both directions from the basic size. If the variation is equal in both directions, then the variation is preceded by a ± symbol. *The plus and minus approach can only be used when the two variations are equal.*

4.5.4 Single Limit Dimensions

When other elements of a feature will determine one limit dimension, MIN or MAX is placed after the other limit dimension. Items such as depth of holes, length of threads, corner radii, and chamfers can use the single limit dimension technique.

4.5.5 Important Terms

Figure 4.29 shows a system of two parts with toleranced dimensions. The two parts are an example of ASME Y14.5M-1994 important terms.

- **Nominal size**—a dimension used to describe the general size, usually expressed in common fractions. The slot in Figure 4.29 has a nominal size of ½″.
- **Basic size**—the theoretical size used as a starting point for the application of tolerances. The basic size of the slot in Figure 4.29 is .500″.

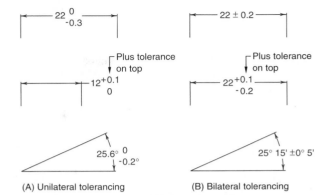

Figure 4.28 Plus and Minus Tolerance System Applied to Various Dimensioning Conditions

- **Actual size**—the measured size of the finished part after machining. In Figure 4.29, the actual size is .501″.
- **Limits**—the maximum and minimum sizes shown by the toleranced dimension. The slot in Figure 4.29 has limits of .502 and .498, and the mating part has limits of .495 and .497. The larger value for each part is the *upper limit,* and the smaller value is the *lower limit.*
- **Allowance**—the minimum clearance or maximum interference between parts, or the *tightest* fit between two mating parts. In Figure 4.29, the allowance is .001, meaning that the tightest fit occurs when the slot is machined to its smallest allowable size of .498 and the mating part is machined to its largest allowable size of .497. The difference between .498 and .497, or .001, is the allowance.
- **Tolerance**—the total allowable variance in a dimension; the difference between the upper and lower limits. The tolerance of the slot in Figure 4.29 is .004″ (.502 − .498 = .004) and the tolerance of the mating part is .002″ (.497 − .495 = .002).
- **Maximum material condition (MMC)**—the condition of a part when it contains the greatest amount of material. The MMC of an external feature, such as a shaft, is the upper limit. The MMC of an internal feature, such as a hole, is the lower limit.
- **Least material condition (LMC)**—the condition of a part when it contains the least amount of material possible. The LMC of an external feature is the lower limit. The LMC of an internal feature is the upper limit.
- **Piece tolerance**—the difference between the upper and lower limits of a single part.
- **System tolerance**—the sum of all the piece tolerances.

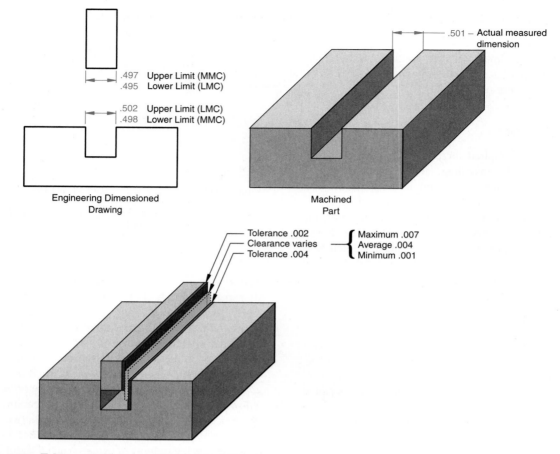

Figure 4.29   **Toleranced Parts and the Important Terms**

4.5.6 Fit Types

The degree of tightness between mating parts is called the *fit*. The basic hole and shaft system shown in Figures 4.30 and 4.31 is an example of the three most common types of fit found in industry.

Clearance fit occurs when two toleranced mating parts will always leave a space or clearance when assembled. In Figure 4.30, the largest that shaft A can be manufactured is .999, and the smallest the hole can be is 1.000. The shaft will always be smaller than the hole, resulting in a minimum clearance of +.001, also called an *allowance*. The maximum clearance occurs when the smallest shaft (.998) is mated with the largest hole (1.001), resulting in a difference of +.003.

Interference fit occurs when two toleranced mating parts will always interfere when assembled. An interference fit *fixes* or *anchors* one part into the other, as though the two parts were one. In Figure 4.30, the smallest that shaft B can be manufactured is 1.002, and the largest the

hole can be manufactured is 1.001. This means that the shaft will always be larger than the hole, and the minimum interference is −.001. The maximum interference would occur when the smallest hole (1.000) is mated with the largest shaft (1.003), resulting in an interference of −.003. To assemble the parts under this condition, it would be necessary to *stretch* the hole or *shrink* the shaft or to use force to press the shaft into the hole. Having an interference is a desirable situation for some design applications. For example, it can be used to fasten two parts together without the use of mechanical fasteners or adhesive.

Transition fit occurs when two toleranced mating parts are sometimes an interference fit and sometimes a clearance fit when assembled. In Figure 4.31, the smallest the shaft can be manufactured is .998, and the largest the hole can be manufactured is 1.001, resulting in a clearance of +.003. The largest the shaft can be manufactured is 1.002, and the smallest the hole can be is 1.000, resulting in an interference of −.002.

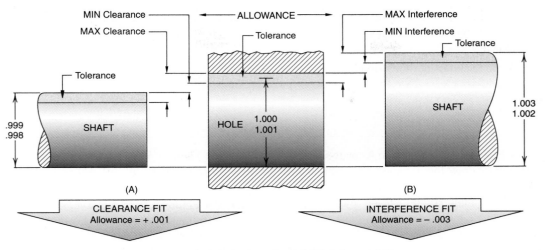

Figure 4.30   **Clearance and Interference Fits between Two Shafts and a Hole**
Shaft A is a clearance fit, and shaft B is an interference fit.

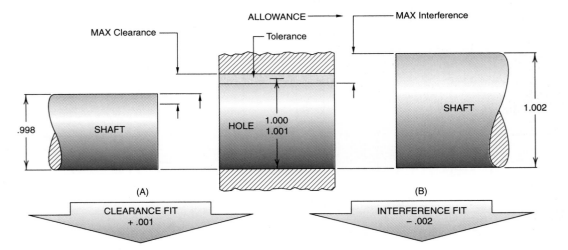

Figure 4.31   **Transition Fit between a Shaft and a Hole**
When the shaft is machined to its smallest diameter (.998), there is a clearance fit with the hole. When the shaft is machined to its largest diameter (1.002), there is an interference fit with the hole.

Questions for Review

1. How are concentric circles best dimensioned?
2. Sketch the symbols for diameter, radius, depth, counterbore, countersink, and square.
3. Where are these symbols placed with respect to their numerical values?
4. When is a small cross required at the center of a radius?
5. When should extension lines be broken?
6. Write a note showing that a .25-inch deep .875-inch diameter hole is to be repeated six times.
7. When is an arc dimensioned with a diameter, and when is one dimensioned with a radius?
8. What is the proper proportion of width to length of arrowheads?

9. What is the difference between limit dimensioning and plus and minus dimensioning?

10. What is MMC?

11. Compare the thickness of dimension lines with object lines.

12. Compare the thickness of dimension lines with extension lines.

Problems

4.1 (Figures 4.32 and 4.33) Using an A- or B-size layout, sketch or draw the assigned problems, using instruments or CAD. Each grid square equals .25″ or 6 mm. Completely dimension the drawing, using one-place millimeter or two-place decimal inches.

4.2 For Figure 4.34 Gusseted Angle Bracket, sketch (or draw with instruments or use a CAD software package) the multiview representation. Include dimensions.

4.3 For Figure 4.35 Angle Clamp, sketch (or draw with instruments or use a CAD software package) the multiview representation. Include dimensions.

4.4 For Figure 4.36 Clevis Eye, sketch (or draw with instruments or use a CAD software package) the multiview representation. Include dimensions.

4.5 For Figure 4.37 Food Mounting, sketch (or draw with instruments or use a CAD software package) the multiview representation. Include dimensions.

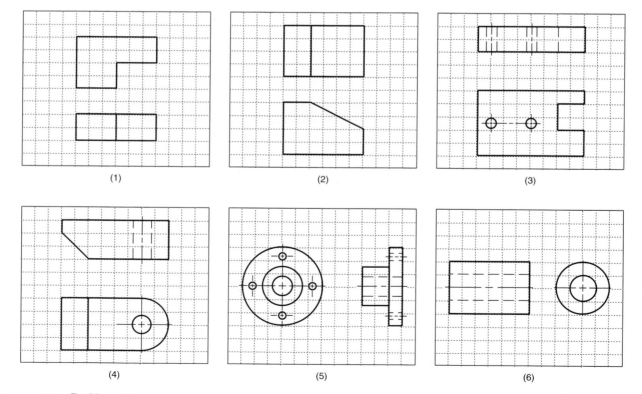

(1)   (2)   (3)

(4)   (5)   (6)

Figure 4.32 **Problem 4.1**

Figure 4.33   Problem 4.1

Figure 4.34   Gusseted Angle Bracket

Figure 4.35 Angle Clamp

Figure 4.36 Clevis Eye

Figure 4.37 Foot Mounting

Chapter

Reading and Constructing Working Drawings

5

OBJECTIVES

After completing this chapter, you will be able to:

1. Define working drawings.
2. Describe how working drawings are used in industry.
3. List the major components of a complete set of working drawings.
4. Describe the differences between detail and assembly drawings.
5. Describe how part numbers, zoning, and tables are used on working drawings.
6. Draw standard representations of threads.
7. Specify a metric or English thread in a note.

5.1 | BASIC CONCEPTS

Engineering drawings are used to communicate designs to others, document design solutions, and communicate design production information. (Figure 5.1) Preceding chapters discussed the development of engineering drawings for the purpose of communicating the design to others. This chapter focuses on communication of the final design for production purposes. These types of drawings are called **working drawings** or **production drawings.**

Figure 5.2 shows the components of the production cycle. **Documenting** is the process of communicating and archiving design and manufacturing information on a product or structure. The documents created include drawings, models, change orders, memos, and reports.

Part of the documenting process includes storing, retrieving, and copying engineering drawings, a process called *reprographics. Archiving* is part of reprographics and involves the storage and retrieval aspects of that process.

Figure 5.1
Engineering drawings are used as a communications tool.

CAD has brought significant changes to this area of design documentation. Using 3-D CAD and modern manufacturing techniques, the need for production drawings is minimized. Rather than creating 2-D drawings of the 3-D model, manufacturers extract production information and geometry directly from the computer model. With both 2-D and 3-D CAD, electronic file storage and distribution eliminate the need for many traditional reprographics activities. With networked computers, personnel in any phase of manufacturing can access the most current version of the production drawings or models on their computer screens.

5.2 | WORKING DRAWINGS

Working drawings are the complete set of standardized drawings specifying the manufacture and assembly of a product based on its design. The complexity of the design determines the number and types of drawings. Working drawings may be on more than one sheet and may contain written instructions called **specifications.**

Working drawings are the *blueprints* used for manufacturing products. Therefore, the set of drawings must: *(a)* completely describe the parts, both visually and dimensionally; *(b)* show the parts in assembly; *(c)* identify all the parts; and *(d)* specify standard parts. The graphics and text information must be sufficiently complete and accurate to manufacture and assemble the product without error.

Generally, a complete set of working drawings for an assembly includes:

1. Detail drawings of each nonstandard part.
2. An assembly or subassembly drawing showing all the standard and nonstandard parts in a single drawing.
3. A bill of materials (BOM).
4. A title block.

5.2.1 Detail Drawings

A **detail drawing** is a dimensioned, multiview drawing of a single part (Figure 5.3), describing the part's shape, size,

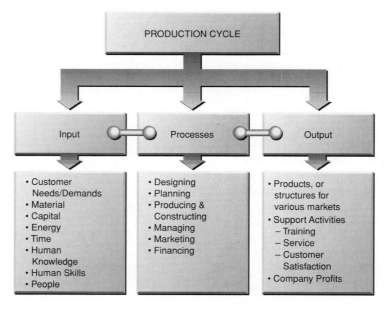

Figure 5.2   Production Cycle
The production cycle includes documenting as a major process.

Figure 5.3   A Detail Drawing of a Single Part Called a Lever

This detail drawing includes three orthographic views, metric dimensions, tolerances, finish information, part number in the title block and on the front view, and material type, as well as other information.

material, and finish, in sufficient detail for the part to be manufactured based on the drawing alone. Detail drawings are produced from design sketches or extracted from 3-D computer models. They adhere strictly to ANSI standards, and the standards for the specific company, for lettering, dimensioning, assigning part numbers, notes, tolerances, etc.

In an assembly, standard parts such as threaded fasteners, bushings, and bearings are not drawn as details, but are shown in the assembly view. Standard parts are not drawn as details because they are normally purchased, not manufactured, for the assembly.

Multiple details on a sheet are usually drawn at the same scale. If different scales are used, they are clearly marked under each detail. Also, when more than one detail is placed on a sheet, the spacing between details is carefully planned, including leaving sufficient room for dimensions and notes. (Figure 5.4) One technique involves lightly blocking in the views for each detail, using construction lines. CAD detail drawings extracted from 3-D models or created in 2-D CAD are more easily manipulated to provide adequate spacing and positioning of multiple details on a single sheet.

5.2.2 Assembly Drawings

An **assembly drawing** shows how each part of a design is put together. (Figure 5.5) If the design depicted is only part of the total assembly, it is referred to as a **subassembly.**

An assembly drawing normally consists of:

1. All the parts, drawn in their operating position.

2. A **parts list** or **bill of materials (BOM)** (shown in Figure 5.5 at the bottom of the drawing sheet, to the left of the title block), showing the detail number for each part, the quantity needed for a single assembly, the description or name of the part, the catalog number if it is a standard part, and the company part number.

3. Leader lines with balloons, assigning each part a **detail number,** in sequential order and keyed to the list of parts in the parts list. For example, the assembly shown in Figure 5.5 has 21 assigned detail numbers, each with a leader and circle, and detail number 5 in the sectioned assembly view is described as an end cap in the parts list.

Figure 5.4   **Blocking in the Views to Determine Spacing Requirements**

Figure 5.5 Multiview Sectioned Assembly Drawing of a Spring Pack

The front view uses variations of section line symbols to distinguish different parts. Courtesy of Kennametal.

4. Machining and assembly operations and critical dimensions related to these functions. In Figure 5.5, the note on the left side view, RECOMMENDED INSTALLATION TORQUE 45–50 Nm (33–37 FT. LBS.), is an assembly operation recommended by the engineer to the person tightening the socket head cap screw.

Dimensions are not shown on assembly drawings, unless necessary to provide overall assembly dimensions, or to assist machining operations necessary for assembly.

Also, hidden lines are omitted in assembly drawings, except when needed for assembly or clarity.

The three basic types of assembly drawings are outline, sectioned, and pictorial.

A **sectioned assembly** gives a general graphic description of the interior shape by passing a cutting plane through all or part of the assembly. (See Figure 5.5) The section assembly is usually a multiview drawing of all the parts, with one view in full section. Other types of sections, such as broken-out and half sections, can also be used.

Chapter 3, "Section and Auxiliary Views," describes the important conventions that must be followed when assemblies are sectioned. These conventions are summarized as follows:

1. Standard parts, such as fasteners, dowels, pins, bearings, and gears, and nonstandard parts, such as shafts, are not sectioned; they are drawn showing all their exterior features. For example, in Figure 5.5, fasteners, such as part number 7, socket head cap screw, are not sectioned.

2. Adjacent parts in section are lined at different angles, using the cast iron or other type of symbol. (Figure 5.5)

3. Thin parts, such as gaskets, are shown solid black.

Sectioned assembly drawings are used for the manufacture and assembly of complicated devices. With CAD, a sectioned assembly can be created by copying detail views and editing them. A 3-D model can also be sectioned to create a sectioned assembly. (Figure 5.6)

A **pictorial assembly** gives a general graphic description of each part, and uses center lines to show how the parts are assembled. (Figure 5.7) The pictorial assembly is normally an isometric view and is used in installation and maintenance manuals.

With 2-D CAD, pictorial assembly drawings can be created using traditional techniques. A 3-D CAD model can also be used to create pictorial assemblies by positioning each part in a pictorial view. (Figure 5.8) Center lines and a parts list are added to complete the drawing. With more sophisticated CAD systems, the part models are referenced from a central CAD database. When specifications on an individual part change, this change is automatically reflected in the assembly model.

5.2.3 Part Numbers

Every part in an assembly is assigned a **part number,** which is usually a string of numbers coded in such a way that a company can keep accurate records of its products. For example, in Figure 5.5, the part with detail number 5 in the sectioned assembly drawing has a company part number of 247987-01, as shown in the parts list. Some assemblies are extremely complicated, with thousands of parts. For example, large aircraft have thousands of parts, and considerable documentation is necessary to design, manufacture, assemble, and maintain the aircraft.

5.2.4 Drawing Numbers

Every drawing used in industry is assigned a number. Each company develops its own standard numbering system, based on various criteria, such as sequential numbers, combinations of numbers and letters, sheet sizes, number of parts in the assembly, model numbers, function, etc. Going back to Figure 5.5, the **drawing number** assigned is 247783R03, which can be found at the lower right-hand corner of the drawing sheet in the bottom of the title block.

5.2.5 Title Blocks

Title blocks are used to record all the important information necessary for the working drawings. The title block is normally located in the lower right corner of the drawing sheet. Figure 5.9 shows the ANSI standard formats for full and continuation title blocks. A continuation title block is used for multiple-sheet drawings and does not contain as much detail as the title block found on the first sheet. Many industries use their own title block formats.

5.2.6 Parts Lists

A complete set of working drawings must include a detailed *parts list* or *bill of material.* Based on ANSI

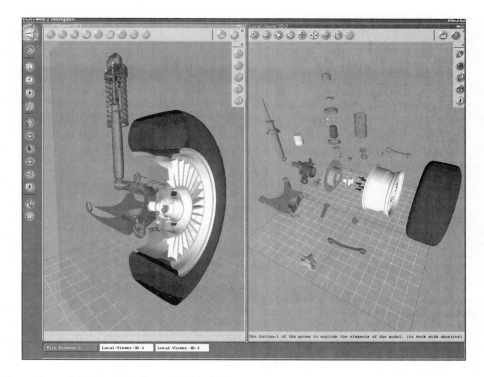

Figure 5.6 **Sectioned Assembly Drawing Created with CAD**
Sectioned assembly drawings are used by assembly technicians to determine how complicated devices are assembled and for design visualization.

Figure 5.7 **Pictorial Assembly Drawing of a Computer Game Joystick**
Courtesy of James Mohler.

Figure 5.8   Pictorial Shaded Exploded Assembly Drawing of a Cordless Screwdriver Made from a 3-D CAD Model
Courtesy of SDRC.

Figure 5.9   ANSI Standard Title Blocks
ANSI Y14.1–1980.

standards, a parts list should be located in the lower right corner above the title block. (Figure 5.10) Additional parts lists may be located at the left of and adjacent to the original block (ANSI Y14.1–1980). As an example, in Figure 5.5, the parts list is located to the left of the title block. A parts list must include a minimum amount of information necessary to manufacture and assemble the part. The information normally included in a parts list is as follows:

1. Name of the part.
2. A detail number for the part in the assembly.
3. The part material, such as cast iron or bronze.
4. The number of times that part is used in the assembly.
5. The company-assigned part number.
6. Other information, such as weight, stock size, etc.

Information on standard parts, such as threaded fasteners, includes the part name and size or catalog number. For example, in Figure 5.11, information on the head liner bushing would be obtained from the tabular drawing from an engineering catalog. Such information includes critical dimensions and model number, such as HL–64–22.

5.2.7 Part Identification

Parts are identified in assembly drawings by a leader line with an arrow that points to the part. The other end of the leader has a **balloon** showing the part detail number. (Figure 5.12) The balloon is approximately four times the height of the number. Part names are given that are placed as close to the part as possible. (Figure 5.13)

5.2.8 Revision Block

Drawing revisions occur because of design changes, tooling changes, customer requests, errors, etc. If a drawing is changed, an accurate record of the change must be created and should contain the date, name of the person making the change, description of the change, the change number, and approval. This information is placed in a **revision**

Figure 5.10   Standard Parts List

The parts list runs vertically for as many rows as are needed to list the parts. ANSI Y14.1–1980.

FLAT WASHERS

USS FLAT WASHERS
Standard washers for general purpose application.
Available in plain or zinc plated steel.
SOLD IN PACKAGE QUANTITIES OF 5 LBS. ONLY

Bolt Size	Diameter I.D.	O.D.	Thk.	Pcs./ Lb.	Plain Steel No.	NET/LB.	Zinc Plated Steel No.	NET/LB.
$3/16$"	$1/4$"	$9/16$"	$3/64$"	303	91081A027	$1.79	91081A027	$1.98
$1/4$"	$5/16$"	$3/4$"	$1/16$"	154	91081A029	1.48	91081A029	1.67
$5/16$"	$3/8$"	$7/8$"	$5/64$"	92	91081A030	1.43	91081A030	1.62
$3/8$"	$7/16$"	1"	$5/64$"	79	91081A031	1.43	91081A031	1.56
$7/16$"	$1/2$"	$1 1/4$"	$5/64$"	41	91081A032	1.37	91081A032	1.54
$1/2$"	$9/16$"	$1 3/8$"	$7/64$"	30	91081A033	1.35	91081A033	1.48
$9/16$"	$5/8$"	$1 1/2$"	$7/64$"	22	91081A034	1.33	91081A034	1.54
$5/8$"	$11/16$"	$1 3/4$"	$9/64$"	13	91081A035	1.35	91081A035	1.46
$3/4$"	$13/16$"	2"	$5/32$"	11	91081A036	1.31	91081A036	1.46
$7/8$"	$15/16$"	$2 1/4$"	$11/64$"	5	91081A037	1.31	91081A037	1.46
1"	$1 1/16$"	$2 1/2$"	$11/64$"	4	91081A038	1.31	91081A038	1.46
$1 1/8$"	$1 1/4$"	$2 3/4$"	$11/64$"	5	91081A039	1.31	91081A039	1.61
$1 1/4$"	$1 3/8$"	3"	$11/64$"	4	91081A040	1.45	91081A040	1.61
$1 3/8$"	$1 1/2$"	$3 1/4$"	$3/16$"	3	91081A042	1.45	91081A042	1.65
$1 1/2$"	$1 5/8$"	$3 1/2$"	$3/16$"	2	91081A041	1.45	91081A041	1.65
$1 5/8$"	$1 3/4$"	$3 3/4$"	$3/16$"	2	91081A043	1.45	91081A043	1.65
$1 3/4$"	$1 7/8$"	4"	$3/16$"	2	91081A044	1.63	91081A044	1.65
2"	$2 1/8$"	$4 1/2$"	$3/16$"	1	91081A046	1.45	91081A046	1.65
$2 1/4$"	$2 3/8$"	$4 3/4$"	$7/32$"	1	91081A047	1.93	91081A047	2.15
$2 1/2$"	$2 5/8$"	5"	$15/64$"	1	91081A048	1.93	91081A048	2.15

Figure 5.11   Typical Page from a Parts Catalog

Part numbers for standard parts are listed in parts catalogs.

Figure 5.12   Balloons

Balloons are used to identify parts by their assigned number in the assembly.

block (Figure 5.14), which is normally in the upper right corner of the drawing, with sufficient space reserved for the block to be expanded downward.

5.2.9 Scale Specifications

The scale used on a set of working drawings is placed in the title block. If more than one scale is used, the scale is shown near the detail drawing. Scales are indicated in

Figure 5.13   Part Name

In detail drawings of an assembly, the part name and detail
number are located near one of the views or in the title block.

metric drawings by a colon, such as 1:2; for the English
system, an equal sign, such as 1=2, is used.

Common English scales found on engineering drawings are:

1=1   Full
1=2   Half
1=4   Quarter
1=8   Eighth
1=10   Tenth
2=1   Double

Common metric scales include:

1:1   Full
1:2   Half
1:5   Fifth
1:10   Tenth
1:20   Twentieth
1:50   Fiftieth
1:100 Hundredth

The designations *METRIC* or *SI* appear in or near the title
block to show that metric dimensions and scale are used
on the drawing.

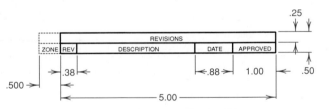

Revision block for A, B, C, and G - sizes

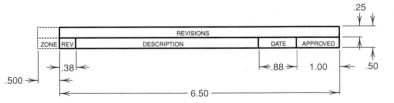

Revision block for D, E, F, H, J, and K - sizes

Figure 5.14   Standard Revision Block

ANSI Y14.1–1980.

5.2.10 Tolerance Specifications

Tolerances are specified in a drawing using toleranced dimensions. For those dimensions that are not specifically toleranced, a **general tolerance note** is used. This note is placed in the lower right corner, near the title block, and usually contains a tolerance and a general statement, such as UNLESS OTHERWISE SPECIFIED.

For example, in Figure 5.15, the general tolerance note specifies the following tolerances for untoleranced dimensions:

One-place decimal inches ±.1
Two-place decimal inches ±.02
Three-place decimal inches ±.005
One-place decimal millimeters ±2.5
Two-place decimal millimeters ±0.51
Three-place decimal millimeters ±0.127

Refer to Chapter 4, "Dimensioning and Tolerancing Practices," for a more detailed explanation of tolerances.

5.2.11 Engineering Change Orders (ECO's)

Making engineering changes after the design has been finalized is very expensive and should be avoided but is necessary if there is a design error, a change request by the customer, or a change in the material or manufacturing process. To make a change in a design most industries require use of an Engineering Change Notice (ECN) or Engineering Change Order. The form used to request a change varies by industry but they all have common features such as:

Identification of what has to be changed in the form of part numbers, part names, and drawing numbers.
An explanation of the need for the requested change.
A list of all documents and departments within an organization that is affected by change.
A description of the change including drawings of the part(s) before and after the change.

Figure 5.15 General Tolerance Note for Inch and Millimeter Dimensions

Courtesy of Kennametal.

Approval of the changes by project managers,
Instructions describing when the changes are to be implemented.

Questions for Review

1. Define working drawings.
2. List the types of drawings commonly found in a complete set of working drawings.
3. List the types of assembly drawings.
4. Define a subassembly.
5. List the important items in a parts list.
6. List the important items in a title block.
7. What is a revision block?

Further Reading

COADE Mechanical Engineering News. COADE Engineering Software, 12777 Jones Rd., Suite 480, Houston, TX 77070, Phone 713–890–4566.

Graves, F. E. "Nuts and Bolts." *Scientific American,* June 1984, pp. 136–144.

Oberg, E.; F. D. Jones; H. L. Horton; H. H. Ryffell. *Machinery Handbook.* 24th ed. New York: Industrial Press, Inc., 1992.

Young, B. "Streamlining the Design Process." *Cadence,* April 1993, pp. 78–80.

Problems

The following figures (Figures 5.16 through 5.22) range from simple to complex mechanical and electromechanical assemblies. For each figure, create a complete set of working drawings. Some of the figures require tolerances to be assigned and missing information to be developed. Some of the assemblies may also be redesigned, to improve on the existing design. For each problem, do the following:

1. Sketch orthographic views of each part, with dimensions. If dimensions are missing, determine what they should be by their relationship to other parts.

2. Determine tolerances as noted or assigned.

3. Determine the scale and size of sheet to use if using traditional tools, or the plot scale and sheet size if using CAD.

4. Determine finished surfaces, and mark them on the sketch.

5. Determine which parts should be represented as section views on the detail drawings.

6. Determine the specifications for all standard parts in the assembly.

7. Create dimensioned detail drawings of each nonstandard part in the assembly.

8. Create an orthographic or exploded pictorial assembly drawing in section.

9. Label all parts in the assembly drawing, using numbers and balloons.

10. Create an ANSI standard parts list with all relevant information for the parts in the assembly.

ITEM	QTY	NAME	DESCRIPTION
1	1 EA.	SUPPORT (LEFT & RIGHT)	STEEL
2	2	ARM	STEEL
3	1	GRIP	STEEL
4	1	PIN A	STEEL
5	1	HINGE A	STEEL
6	1	HINGE B	STEEL
7	1	SHAFT	STEEL
8	1	HANDLE	STEEL
9	3	PIN B	STEEL
10	1	PIN C	STEEL

QUICK ACTING HOLD-DOWN CLAMP

Figure 5.16　Quick Acting Hold-Down Clamp

R .3125
.250
R .10
R .10
R .1562
.250
.2187
45°
6X Ø .250
.750
.500
1.250
.125
1.250
.8750 .375
.375
.375
R .25
.375
1.250
1.000
2.500
.125

PART NAME: SUPPORT RIGHT & LEFT NEEDED

Ø .2343
Ø .750
R .0312 FILLET BOTH SIDES
Ø .250 BOTH SIDES
PEEN AT ASSEMBLY BOTH ENDS
.750
.1875 BOTH SIDES

PART NAME: HINGE A

1.125
Ø .3125
Ø .250
PEEN AT ASSEMBLY BOTH ENDS
.750
.1875

PART NAME: PIN B

Ø .3125
Ø .250
.375
PEEN AT ASSEMBLY BOTH ENDS
1.125
.375

PART NAME: PIN C

Ø .2812
Ø .250
PEEN AT ASSEMBLY BOTH ENDS
.750
2X .1875

PART NAME: PIN A

Ø .3125
R .03125 ALL EDGES
.750
.375
.250
.375
1.375

PART NAME: GRIP

R .3125
1.375
2.250
.125
R .25
4°
R .3125
3X Ø .250
R .500

PART NAME: ARM

Figure 5.16 Continued

.375-16UNC

Ø .700

R .0312 FILLET
BOTH ENDS

Ø .3125
BOTH ENDS

PEEN AT ASSEMBLY
BOTH SIDES

1.000

.125
BOTH SIDES

PART NAME: HINGE B

Ø .6250

Ø .266
CENTERED ON
FEATURE

.375-16UNC

.4375

4.1875

5.0625

Ø .250

PART NAME: SHAFT

Ø .250

3.000

PART NAME: HANDLE

Figure 5.16   Continued

ITEM	QTY	NAME	DESCRIPTION
1	2	BUSHING	BRONZE
2	1	BASE	STEEL
3	1	CAP	STEEL
4	6	HEX HEAD BOLT	.50-13UNC X 2.00

DOUBLE BEARING

Figure 5.17 Double Bearing

PART NAME: CAP

PART NAME: BUSHING

PART NAME: BASE

Figure 5.17   Continued

Figure 5.18   Double V-Block

SOCKET HEAD SHOULDER SCREW 5/8" DIAMETER
2.50" SHOULDER LENGTH RC1 STEEL

BUSHING
OUTER DIAMETER 1.00" FN2
INNER DIAMETER 5/8" RC1
THICKNESS 1/2"
BRONZE

WHEEL
DIAMETER 4.00"
WIDTH 1.00" BODY .50" THICK
HUB WIDTH 1.50" X .75" O.D.
GROOVE 45° AND CENTERED
CAST IRON

.50

DETAIL WHEEL

FN2 FIT

WASHER
INNER DIAMETER 5/8"
BRONZE

SUPPORT COLD
ROLLED STEEL

HEX HEAD NUT
1/2" - 13NC

HEX HEAD BOLT
.500–13UNC X 1.00

.50

R .625

Ø 3.25

2.00

1.50

4.00

R .125

1.50

1.00

2.00

4.00

4.00

4X .500-13UNC

11.50

6.50

2.50

.75

1.00

3.00

BASE
COLD ROLLED STEEL

WHEEL SUPPORT

Figure 5.19 Wheel Support

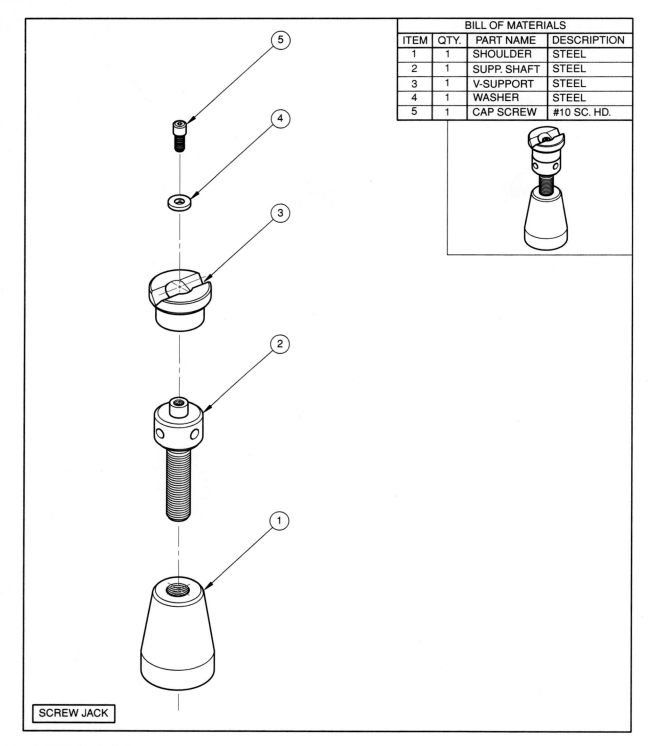

BILL OF MATERIALS			
ITEM	QTY.	PART NAME	DESCRIPTION
1	1	SHOULDER	STEEL
2	1	SUPP. SHAFT	STEEL
3	1	V-SUPPORT	STEEL
4	1	WASHER	STEEL
5	1	CAP SCREW	#10 SC. HD.

SCREW JACK

Figure 5.20 Screw Jack

#10-32UNF X .578 DP

Ø .3750

.2660

3X .031 X 45°

Ø 1.00

.6251

.1875

Ø .218
2 HOLES 90° APART
CENTER .5785 FROM TOP
OF PART

2.7656

1.6875

.500 - 20 - UNF - 3A

PART NAME: SUPPORT SHAFT

Ø .406
⌴ Ø .5625
▽ .5625

3X .031 X 45°

90° V CUT
▽ .218

Ø 1.250

.8125   .563

Ø 1.00

PART NAME: V-SUPPORT

10° OFF OF VERTICAL TAPER

Ø .453
▽ 1.25
.500-20UNF
▽ 1.25
FROM BOTTOM
Ø .625
▽ 1.094
⌴ Ø 1.125
▽ .094

R .0625 FILLET

Ø 1.00

Ø .203

Ø .500

.0625

2.00

.500

PART NAME: WASHER

.0625 X 45°

Ø 1.50

PART NAME: SHOULDER

Figure 5.20   Continued

ITEM	QTY	NAME	DESCRIPTION
1	1	BODY	CAST IRON
2	2	ROUND HEAD MACHINE SCREW	#4-48UNF X .250
3	1	PLATE	ALUMINIUM
4	1	SHAFT	STEEL
5	1	WOODRUFF KEY	
6	1	RETAINER	STEEL
7	3	ROUND HEAD MACHINE SCREW	#10-32UNF X .500
8	1	ARM	STEEL
9	1	HEX HEAD NUT	.375-24UNF

BUTTERFLY VALVE

Figure 5.21 Valve

PART NAME: BODY

PART NAME: RETAINER

PART NAME: ARM

PART NAME: SHAFT

PART NAME: PLATE
(ELLIPTICAL)

Figure 5.21 Continued

BILL OF MATERIALS			
ITEM	QTY.	PART NAME	DESCRIPTION
1	1	INDEX	STEEL
2	1	INDEX PLATE	STEEL
3	1	GENEVA	STEEL
4	1	BUSHING	BRASS
5	2	WASHER	STEEL
6	1	FLAT HEX NUT	STEEL
7	1	CAP SCREW	#3 SC. HD.
8	2	SHAFT	STEEL
9	1	SET SCREW	STEEL

GENEVA CAM

Figure 5.22   Geneva Cam

PART NAME: INDEX PLATE

PART NAME: SHAFT

PART NAME: INDEX

PART NAME: BUSHING

PART NAME: WASHER

PART NAME: GENEVA

Figure 5.22   Continued

Problems Refer to Figures 5.16 through 5.22

Sketch the necessary views, then add dimensions.

Sketch Number: _____

Name:_____

Div/Sec: _____

Date:_____

Problems Refer to Figures 5.16 through 5.22

Sketch the necessary views, then add dimensions.

Sketch Number: _____

Name:_____

Div/Sec: _____

Date:_____

Problems Refer to Figures 5.16 through 5.22

Sketch the necessary views, then add dimensions.

Sketch Number: _____

Name:_____

Div/Sec: _____

Date:_____

Problems Refer to Figures 5.16 through 5.22

Sketch the necessary views, then add dimensions.

Sketch Number: _____

Name:_____

Div/Sec: _____

Date:_____

Problems Refer to Figures 5.16 through 5.22

Sketch the necessary views, then add dimensions.

Sketch Number: _____

Name:_____

Div/Sec: _____

Date:_____

Design and 3-D Modeling

6

OBJECTIVES

After completing this chapter, you will be able to:

1. Describe the engineering design process and the role graphics plays.
2. Describe concurrent engineering and design for manufacturability (DFM).
3. Describe the types of modeling systems used in 3-D modeling.
4. Explain the role 3-D modeling plays in the engineering design process.
5. List and describe the modeling techniques used in design.
6. Describe the rapid prototyping process.
7. List and describe the analysis techniques used in design.
8. Explain how graphing and visualization can be used in the design process.
9. Describe the total quality management (TQM) process.

6.1 | ENGINEERING DESIGN

Design is the process of conceiving or inventing ideas mentally and communicating those ideas to others in a form that is easily understood. Most often the communications tool is graphics.

Engineering design is one of the processes normally associated with the entire business or enterprise, from receipt of the order or product idea, to maintenance of the product, and all stages in between. (Figure 6.1) The design process requires input from such areas as customer needs, materials, capital, energy, time requirements, and human knowledge and skills.

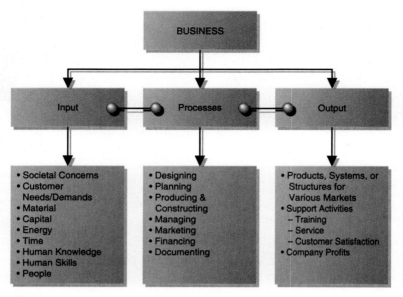

Figure 6.1 The Business Process
A manufacturing business or enterprise includes all the inputs, processes, and outputs necessary to produce a product or construct a structure. Designing is one of the major processes in such a business.

6.1.1 Traditional Engineering Design

Traditional engineering design is a linear approach divided into a number of steps. For example, a six-step process might be divided into problem identification, preliminary ideas, refinement, analysis, documentation, and implementation. The design process moves through each step in a sequential manner; however, if problems are encountered, the process may return to a previous step. This repetitive action is called **iteration** or looping. Many industries use the traditional engineering design process; however, a new process is developing that combines some features of the traditional process with a team approach that involves all segments of a business.

6.1.2 Concurrent Engineering Design

The **production process** executes the final results of the design process to produce a product or system. In the past, the creative design process was separated from the production process. With the advent of computer modeling, this separation is no longer necessary, and the modern engineering design approach brings both processes together.

Concurrent engineering is a nonlinear team approach to design that brings together the input, processes, and output elements necessary to produce a product. The people and processes are brought together at the very beginning, which is not normally done in the linear approach. The team consists of design and production engineers, technicians, marketing and finance personnel, planners, and managers, who work together to solve a problem and produce a product. Many companies are finding that concurrent engineering practices result in a better, higher-quality product, more satisfied customers, fewer manufacturing problems, and a shorter cycle time between design initiation and final production.

Figures 6.2 and 6.3 represent the concurrent approach to engineering design, based on 3-D modeling. The three intersecting circles represent the concurrent nature of this design approach. For example, in the ideation phase, design engineers interact with service technicians to ensure that the product will be easily serviceable by the consumer or technician. This type of interaction results in a better product for the consumer. The three intersecting circles also represent the three activities that are a major part of the concurrent engineering design process: ideation, refinement, and implementation. These three activities are further divided into smaller segments, as shown by the items surrounding the three circles.

The center area in Figure 6.3 represents the 3-D computer model and reflects the central importance of 3-D modeling and graphics knowledge in engineering design and production. With the use of a modeling approach, everyone on the team can have access to the current design through a computer terminal. This data sharing is critically important to the success of the design process.

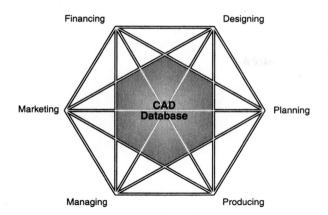

Figure 6.2 **Sharing the CAD Database**
The concurrent engineering model shows how every area in an enterprise is related, and the CAD database is the common thread of information between each area.

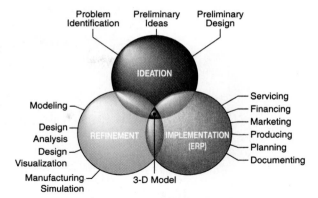

Figure 6.3 **Concurrent Engineering Design**
The engineering design process consists of three overlapping areas: ideation, refinement, and implementation, which all share the same 3-D CAD database.

Through the sharing of information, often in the form of a database, it is possible for all areas of the enterprise to work simultaneously on their particular needs as the product is being developed. For example, a preliminary 3-D model could be created by the design engineers early in the ideation phase. A mechanical engineer could use the same 3-D model to analyze its thermal properties. The information gained from this preliminary analysis could then be given to the design engineers, who could then make any necessary changes early in the ideation phase, minimizing costly changes later in the design process.

6.2 | 3-D MODELING

Traditionally, the means of communication in the engineering design process was through paper drawings done by hand. With the increasing availability of CAD tools, these 2-D technical drawings are being produced on computer. More recently, 3-D modeling software has become available on increasingly powerful PCs and increasingly inexpensive engineering workstations. (Figure 6.4) Because 3-D modeling systems create models of the product being designed, this system offers considerably more possibilities as to how it can be integrated into the design process than a 2-D CAD drawing does.

The following section offers an overview of the most common approaches for generating 3-D computer models. Because working with a 3-D model is so much different than creating a 2-D drawing, the ways in which these models are

constructed and viewed are also discussed. As the various stages of the design process are presented, the ways in which 3-D modeling can play an integral role will be outlined.

6.2.1 Wireframe Modeling

The simplest 3-D modeler is a **wireframe modeler.** In this type of modeler, which is a natural outgrowth of 2-D CAD, two types of elements must be defined in the database: *edges* and *vertices.* (Figure 6.5) For the tetrahedron in the figure, the vertex list contains the geometric information on the model. Each vertex is defined by an (X, Y, Z) coordinate, which anchors the model in space. The topology of the model is represented by the edge list. The edge list does not contain coordinate information. The location, orientation, and length of an edge must be derived indirectly, through calculations of the vertices at either end of the edge. For example, edge E1 consists of vertices V1 and V2. The coordinate locations of V1 (0,0,0) and V2 (1,0,0) indicate that E1 has a length of 1 and is oriented along the X axis.

6.2.2 Surface Modeling

Surface models define the surface features, as well as the edges, of objects.

Different types of spline curves are used to create surface patches with different modeling characteristics. For example, the advantage of Bezier surface patches is that they are easy-to-sculpt natural surfaces. (Figure 6.6) The control points are an intuitive tool with which the user can work.

In contrast, B-spline patches allow local control; moving one control point does not affect the whole surface. With B-splines, it is much easier to create surfaces through predefined

Figure 6.4 UNIX Workstation
UNIX workstations are popular for particularly demanding 3-D modeling work. Courtesy of Dassault Systems.

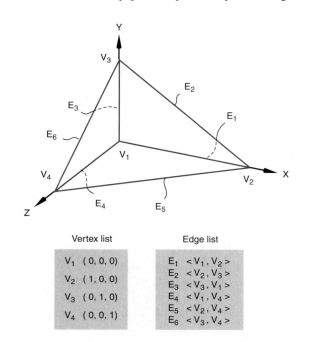

Vertex list

V_1	$(0, 0, 0)$
V_2	$(1, 0, 0)$
V_3	$(0, 1, 0)$
V_4	$(0, 0, 1)$

Edge list

E_1	$<V_1, V_2>$
E_2	$<V_2, V_3>$
E_3	$<V_3, V_1>$
E_4	$<V_1, V_4>$
E_5	$<V_2, V_4>$
E_6	$<V_3, V_4>$

Figure 6.5 The Vertex and Edge List of a Wireframe Model

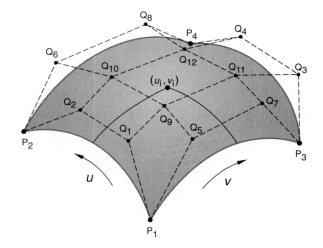

Figure 6.6 A Bezier Bicubic Surface Patch
The patch consists of four connected Bezier curves and 16 control points.

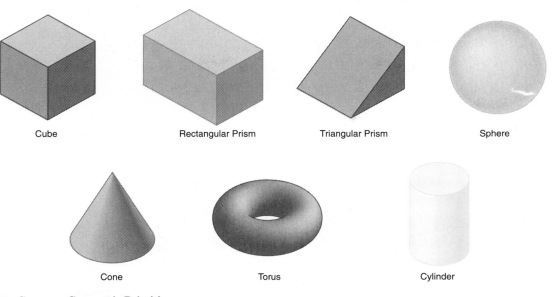

Figure 6.7   Common Geometric Primitives

points or curves. NURBS surfaces use *rational* B-splines, which include a *weighting* value at each point on the surface. The weighting value allows some points to have more influence over the shape of the curve than other points. This means that a wider variety of curved surfaces are possible than with regular B-splines. Because NURBS surfaces can also precisely describe conic surfaces, they are gaining popularity in many tasks previously handled by other types of 3-D modelers.

6.2.3 Solid Modeling

Through the use of parametric description, surface modelers can accurately describe the surface of an object. Often, however, information about the inside of an object, that is, its solidity, is also required. **Solid models** include volumetric information, that is, what is on the inside of the 3-D model, as well as information about the surface of an object. In this case, the surface of the model represents the boundary between the inside and outside of the object.

Primitive Modeling   Many objects, including most mechanical parts, can be described mathematically using basic geometric forms. Modelers are designed to support a set of *geometric primitives,* such as cubes, right rectilinear prisms (i.e., blocks), right triangular prisms (i.e., wedges), spheres, cones, tori, and cylinders. (Figure 6.7) Although most geometric primitives have unique topologies, some differ only in their geometry, like the cube and the right rectilinear prism.

Many modelers also allow primitives to be joined together to create more complex objects. The user mentally decom-

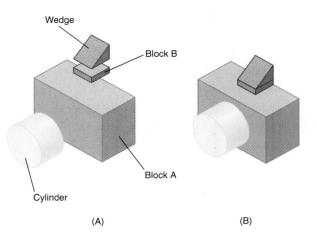

Figure 6.8   A Camera Described with Geometric Primitives
Additive modeling with geometric primitives allows a variety of objects to be represented.

poses the object into a collection of geometric primitives and then constructs the model from these elements. (Figure 6.8)

There are a number of positive aspects to using a limited set of primitives. First, limiting the allowed topologies reduces the risk of an *invalid* model. Typically, the only safeguards needed are limits or rules on parameter input. For example, the diameter of a cylinder could not be specified as zero. Second, the limited number of allowable shapes means a very concise and efficient database. Third, the limited range of primitives means a model that is unique and is easy to validate.

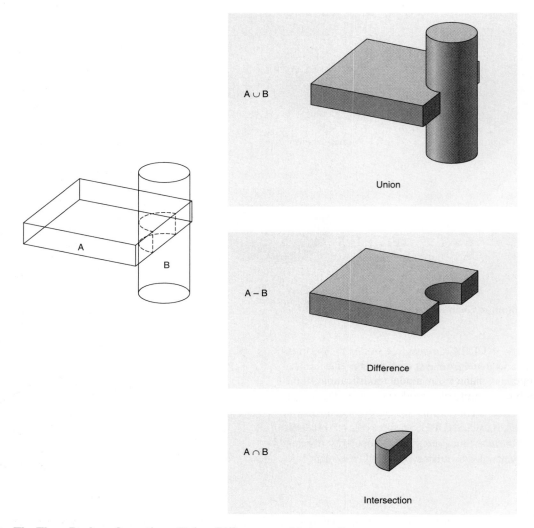

Figure 6.9 The Three Boolean Operations: Union, Difference, and Intersection
The three operations, using the same primitives in the same locations, create very different objects.

Constructive Solid Geometry (CSG) Modeling **Constructive solid geometry (CSG)** modeling is a powerful technique that allows flexibility in both the way primitives are defined and the way they are combined. The relationships between the primitives are defined with **Boolean operations.** There are three types of Boolean operations: **union** (∪), **difference** (–), and **intersection** (∩). Figure 6.9 shows how each of these operations can be used to create different forms. The critical area is the place where two objects overlap. This is where the differences between the Boolean operations are evident. The union operation is essentially additive, with the two primitives being combined. However, in the final form, the volume where the two primitives overlap is only represented once. Otherwise there would be twice as much material in the area of overlap, which is not possible in a real object. With a difference operation, the area of overlap is not represented at all. The final form resembles one of the original primitives with the area of overlap removed. With the intersection operation, *only* the area of overlap remains; the rest of the primitive volume is removed.

Boundary Representation (B-Rep) Modeling **Boundary representation (B-rep) modeling** and CSG modeling are the two most popular forms of solid modeling. With CSG modeling, surfaces are represented indirectly through half-spaces; with B-rep modeling, the surfaces, or *faces,* are themselves the basis for defining the solid. The face of a B-rep model is fundamentally different from a face in a wireframe modeler. Although both include a set of edges connected at vertices, a B-rep face explicitly represents an oriented surface. There are two sides to this surface: one is on the *inside* of the object (the solid side), and the other is on the *outside* of the object (the void side).

6.3 | CONSTRAINT-BASED MODELING

When working with a traditional 3-D modeling system, defining the size or shape of a geometric feature is typically done independently of any other feature. It is completely up to the user of the CAD system to make sure the resulting model reflects the real-world design constraints the final product will have to face. With a simple part consisting of three overall dimensions and one or two features (e.g., holes or slots), making sure these features are properly placed as the part is constructed is not a major task. On the other hand, with complex parts or large assemblies, the total number of features to track can be a major task. Traditionally with CAD systems, it has been up to the operator to ensure that when one part is modified, related features on other parts are also updated. **Constraint-based modeling** is a technique that can help the CAD operator manage the model modification process.

Although the use of solid modeling grew steadily during the 1980s, many of the productivity gains promised by CAD vendors were not being realized by companies. One of the reasons was that the process of creating a solid model was much more abstract than the process of designing real-world products.

In a constraint-based modeler, a part is created by describing the relationship of geometric elements with equations and logical relationships. For example, in a traditional CAD modeler, if a plate is to have a length equal to 32 mm and a width of half the length, a rectangle 32 mm long and 16 mm wide would be created. (Figure 6.10A) In a constraint-based modeler, the geometric relationships are coded directly into the model with *parameters:* the length would be defined as $P_1 = 32$ mm and the width defined as $P_2 = P_1/2$. (Figure 6.10B) With the geometry defined with parameters, the equations are solved to derive dimensional values for each of the features. The power of this approach is seen when the model is modified. Instead of having to individually update all related dimensions, one dimension can be altered, and all dimensions linked through parameters automatically reflect the change. For example, if the length of the plate is reduced ($P_1 = 20$), the width automatically reduces by the appropriate amount to reflect the change in length. (Figure 6.10C)

Like dimensions, parameters can be associated with geometric features such as lines, curves, and planes. Un-

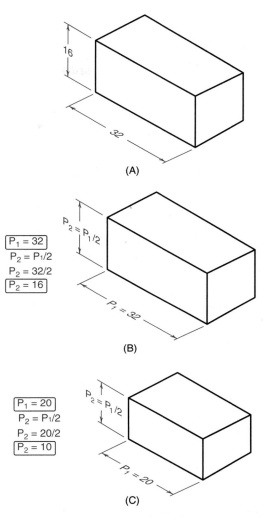

Figure 6.10   Traditional and Parametric Part Definition
(A) A traditional CAD modeler creates geometry of a specified size. (B) With a constraint-based modeler, geometry is defined through parameters represented as an equation. (C) Parameters can easily be altered to represent new values.

like dimensions, parameters do not have to represent a single static value. Parameters can be

- Assigned numeric values (e.g., $P_1 = 32$ mm).
- Related to other parameters through equations (e.g., $P_2 = P_1 + 20$).
- Related to other parameters through geometric relationships (e.g., P_3 is parallel to P_4).
- Varied based on logical relationships (e.g., IF $P_5 > 24$ mm, THEN $P_6 = 6$ mm, ELSE $P_6 = 4$ mm).

It is important to remember that parameters are always related back to geometric features. A parameter that represents

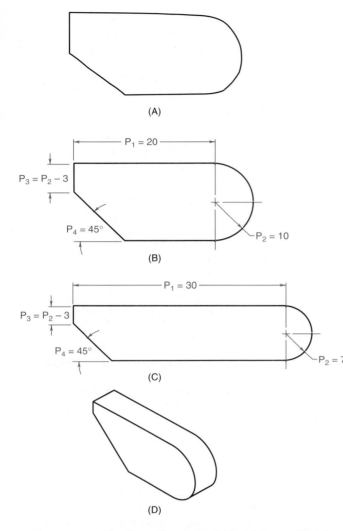

Figure 6.11 Creating a Constrained Model from a Sketch Profile

(A) A rough sketch defines the basic geometry of the 2-D profile. (B) Constraints are added to define the size and shape of the profile. (C) The size and shape of the profile can be altered at any time. (D) Once constrained, the profile can be extruded or swept into a 3-D part.

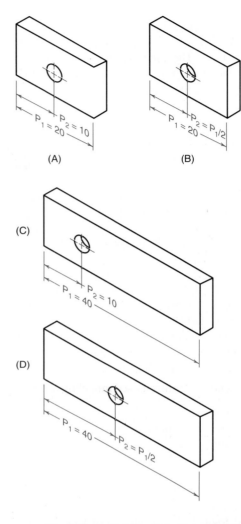

Figure 6.12 The Effect of Design Intent on Model Changes

(A) Though these two parts are initially the same, the parametric constraints represent two different design intents. The part on the left intends to have the hole fixed at an offset of 10 from the left edge whereas the part on the right intends to have the hole centered. (B) The differences in the design intent can be seen when the overall width of the part is altered. Only the hole on the part on the right shifted to the center.

a numeric value measures a distance or angle between two features (i.e., points or lines). Likewise, geometric relationships such as parallelism are made between features such as lines or planes.

In many constraint-based modelers, the modeling process begins by creating a 2-D *sketch profile*. (Figure 6.11A) Unlike in a traditional 2-D or 3-D CAD package, the geometry does not have to be created with a high degree of accuracy; it only needs to represent the basic geo-

metric features. The next step is to *constrain* the model by assigning enough parameters to fully define the size and shape of the 2-D profile. (Figure 6.11B) Depending on the modeler, some of these parameters may be automatically assigned. For example, if the two horizontal elements are within a preset tolerance range (say, 5 degrees) of parallel, the two edges are constrained to be parallel. You might think of this as being like a snapping or grid system, but, in fact, it is much more powerful. As the part is resized, the sides will stay parallel no matter how far apart they are. Because all of the features are parametrically defined, one or more of the parametric values can be altered, and all related parameters will automatically update. (Figure 6.11C) Once the sketch profile has been constrained and the parameters assigned appropriate values, a 3-D part can be created by an extrusion or sweep operation. (Figure 6.11D)

In Figure 6.12A, two plates are created with holes and two different design intents. In the plate on the left, the hole was intended to be placed 10 mm from the left edge, while the plate on the right had the hole placed in the center. When the overall width is set to 20 mm, no difference is seen in the two models, but when the overall width is set to a different value, the difference in design intent is immediately seen. (Figure 6.12B) With a traditional modeler, an engineer viewing the model in Figure 6.12A would not be able to tell what the design intent was and, therefore, how the hole should shift if the model was altered.

6.4 | FEATURE-BASED MODELING

Another important advancement in 3-D solids modeling is the introduction of **feature-based modeling.** Like constraint-based modeling, feature-based modeling is an attempt to make modeling a more efficient process more in tune with how designers and engineers actually work. Feature-based modeling bundles commands together to automate the process of creating and modifying features that represent common manufacturing operations. Usually implemented in modelers that also have constraint capabilities, feature-based modeling systems use special dialog boxes or other interface elements that allow users to input all of the variables needed to create a common manufactured feature. (Figure 6.13)

Examples of manufactured features created through special feature-based dialog boxes include the following:

• Blind holes.
• Counterbores and countersinks.
• Slots.
• Bosses.

The hole dialog box shown in Figure 6.13 is a good example of automating the process of creating features in a model. The feature is broken down into its essential variables, with each variable represented by an input in the dialog box. Variables such as the hole's diameter have a value typed in, while the depth can be set to "through" by clicking a button or set to a finite value.

The variables entered through the dialog box largely define the *shape* and *size*. Once these variables of the feature are

Figure 6.13   Example of a Dialog Box in a Feature-Based Modeler
Notice that all of the essential variables defining the feature are represented as inputs in the dialog box. Typically, many of these variables can also be defined as parameters linked to other features on the model.

defined, the *location* is defined. By convention, features usually don't constitute an entire part. For that reason, the feature is typically located somewhere on a *base part.* A feature such as a blind hole is located by indicating its orientation to a face and distance from two edges. (Figure 6.14) In a constraint-based modeler, all of the variables of the feature—its shape, size, and location—are parametrically controlled and can be updated at any time. In addition, the parameters defining the feature can also be linked to other parameters defining the part. So, for example, the depth of a hole might be related to the overall thickness of the base part.

Earlier in the chapter, Boolean operations were introduced as the primary means of modifying a solids model. With feature-based modeling, one or more Boolean operations are manipulated using intelligent dialog boxes to create common features such as holes, slots, and bosses. Feature- and constraint-based techniques together give the user the ability to easily create and modify common manufactured features. When combined with parts libraries of common fasteners and other parts, assemblies using standard components and manufacturing techniques can be quickly modeled.

Figure 6.14 Locating a Feature on the Base Part

The blind counterbore is located on the base part relative to a face on the part. The overall depth of the counterbore is linked to the overall height of the part through a parametric equation.

6.5 | 3-D MODELING AND THE DESIGN PROCESS

CAD was introduced into most businesses as an automated drafting tool and was then defined as computer-aided drafting. The introduction of 3-D modeling systems has transformed CAD into computer-aided *design.*

CAD 3-D modeling plays an important role in many newly emerging manufacturing techniques, including computer-aided manufacturing (CAM), computer-integrated manufacturing (CIM), concurrent engineering, and design for manufacturability (DFM). All of these manufacturing techniques are aimed at shortening the design cycle, minimizing material and labor expenditures, raising product quality, and lowering the cost of the final product. Central to these goals is better communications within a company. By sharing the 3-D database of the proposed product, more people can be working simultaneously on various aspects of the design problem. The graphics nature of the database has convinced many that 3-D modeling is a superior method of communicating many of the design intents.

Questions for Review

1. Describe the design process.
2. Define the three types of Boolean operations and sketch examples of each one. Can you derive the same final object using different Boolean operations and/or primitives?
3. Describe feature-based modeling.
4. Describe constraint-based modeling.

Further Reading

Anand, V. *Computer Graphics and Geometric Modeling for Engineers.* New York: John Wiley, 1993.

Beakley, G. C., and H. W. Leach. *Introduction to Engineering Design Graphics.* New York: Macmillan, 1973.

Bolluyt, J., M. Stewart, and A. Oladipupo. *Modeling for Design Using SilverScreen.* Boston: PWS-Kent, 1993.

Chase, R. B., and N. J. Aquilano. *Production and Operations Management.* 6th ed. Homewood, IL: Richard D. Irwin, 1992.

Dieter, G. E. *Engineering Design: A Materials and Process Approach.* New York: McGraw-Hill, Inc., 1983.

Dreyfuss, H. *The Measure of Man, Human Factors in Design.* New York: Whitney Library of Design, 1967.

Foley, J. D., et al. *Computer Graphics: Principles and Practice.* Reading, MA: Addison-Wesley, 1992.

Guide for Patent Draftsmen (Selected Rules of Practice Relating to Patent Drawings). Washington, DC: U.S. Department of Commerce, Patent and Trademark Office, 1989.

Hunt, V. D. *Quality in America.* Homewood, IL: Business One Irwin, 1992.

LaCourse, D. E., ed. *Handbook of Solid Modeling.* New York: McGraw-Hill, 1995.

McCarthy, E. J., and W. E. Perreault, Jr. *Essentials of Marketing.* 5th ed. Homewood, IL: Richard D. Irwin, 1991.

Mortenson, M. *Geometric Modeling.* New York: John Wiley, 1985.

Oberg, E., F. D. Jones, H. L. Horton, and H. H. Ryffell. *Machinery's Handbook.* 24th ed. New York: Industrial Press, 1992.

Pillers, M. "Using AME for New Product Development." *CADENCE,* May 1993, pp. 45–48.

Taylor, D. L. *Computer-Aided Design.* Reading, MA: Addison-Wesley, 1992.

Zeid, I. *CAD/CAM Theory and Practice.* New York: McGraw-Hill, 1991.

3-D Modeling and Design Project
3-D Modeling Projects

One of the best ways to learn engineering design graphics principles and to use 3-D modeling tools is to use a real object to simulate the process. When the author first learned CAD, he used this technique using a road bike. Each chapter in the textbook will have a special problem related to creating a 3-D model of an assembly. There are a number of inexpensive products that could be used for this exercise that can easily to purchased at stores, such as a disposable camera, mechanical pencil, or various tools. For this exercise you will be using a stapler like that shown in Figure 1A. The stapler in the figure is a Swingline Desktop Stapler #81800 that can be purchased at popular retail stores, such as Walmart, Kmart, and Staples. To avoid having curved surfaces, Swingline Desktop Stapler #74700 can be used. You will be assigned to teams to purchase a stapler, disassemble the item, measure the parts, create the design sketches, make simple design changes, model the parts, create working drawings for manufacturing, perform simple assembly analyses, and create rendered models of the assemblies.

To start on this assignment the stapler must be reversed engineered by disassembling the stapler and measuring all the parts using a caliper and a micrometer. As each part is measured, design sketches are made of each part with dimensions.

Figure 1A   Desktop stapler for design project

Figure 1B   Parts to be sketched

Problems

6.1 In Figure 6.15, there are 12 objects swept, using 12 different profiles. Match the objects with the same profile used to create 3-D objects.

6.2 (Figures 6.16–6.22) Assign different Boolean operations to the seven assembled primitive parts, then do the following:

a. Sketch the resulting composite solids.

b. Use a solid modeling software to create the primitives with the given dimensions, then verify your sketches by performing the same Boolean operations on the computer.

6.3 (Figures 6.23–6.24) Using the given information for feature-based modeling, do the following:

a. Using a scale assigned by your instructor, measure the profiles and workpiece. On isometric grid paper, sketch the resulting workpiece after the feature-based modeling is performed.

b. Do the same operations with CAD and compare the results with your sketch.

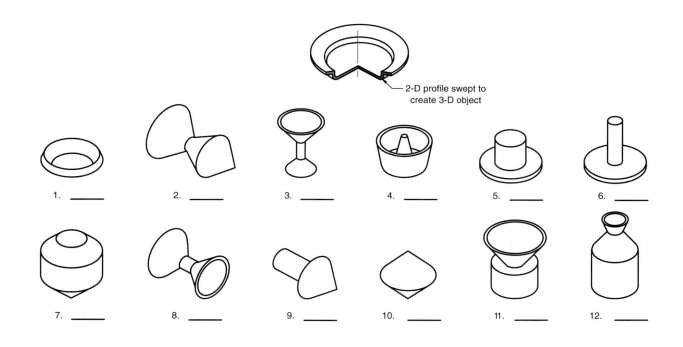

1. _____   2. _____   3. _____   4. _____   5. _____   6. _____

7. _____   8. _____   9. _____   10. _____   11. _____   12. _____

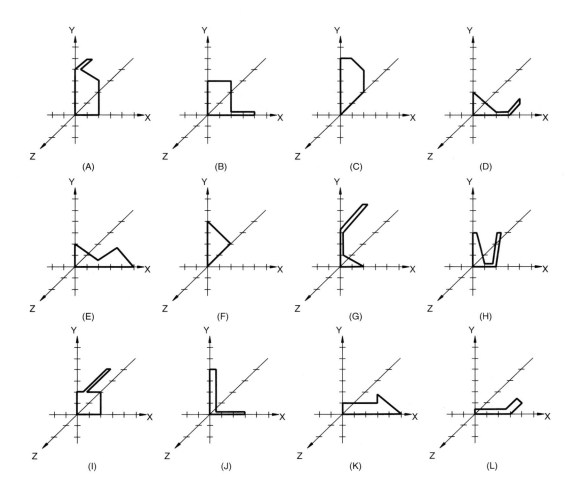

Figure 6.15   **Match 2-D Profiles to 3-D Objects**

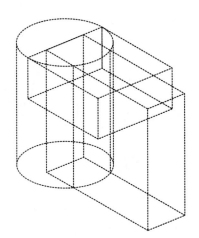

Figure 6.16   Assembled Primitive Parts for Boolean Operations

Figure 6.17

Figure 6.18

Figure 6.19

Figure 6.20

Figure 6.21

Figure 6.22

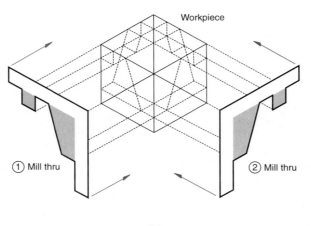

① Mill thru   ② Mill thru

(A)

① Mill thru

(B)

Figure 6.23   Feature-Based Modeling Information

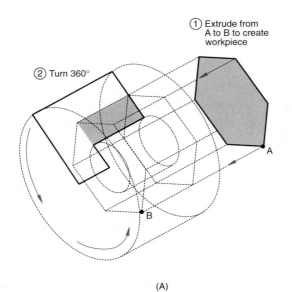

① Extrude from A to B to create workpiece

② Turn 360°

(A)

② Bore

Workpiece

① Mill thru 15° to the horizontal

(B)

Figure 6.24

Problem 6.3

Refer to Figures 6.23 and 6.24. Using a scale assigned by your instructor, measure the profiles and workpiece, then sketch the resulting workpiece after the feature-based modeling is completed.

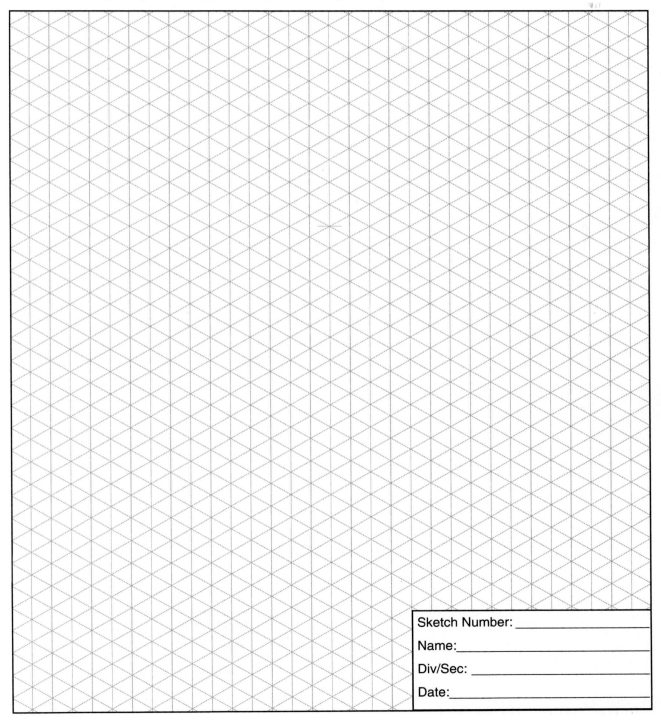

Sketch Number: _____

Name:_____

Div/Sec: _____

Date:_____

Problem 6.3

Refer to Figures 6.23 and 6.24. Using a scale assigned by your instructor, measure the profiles and workpiece, then sketch the resulting workpiece after the feature-based modeling is completed.

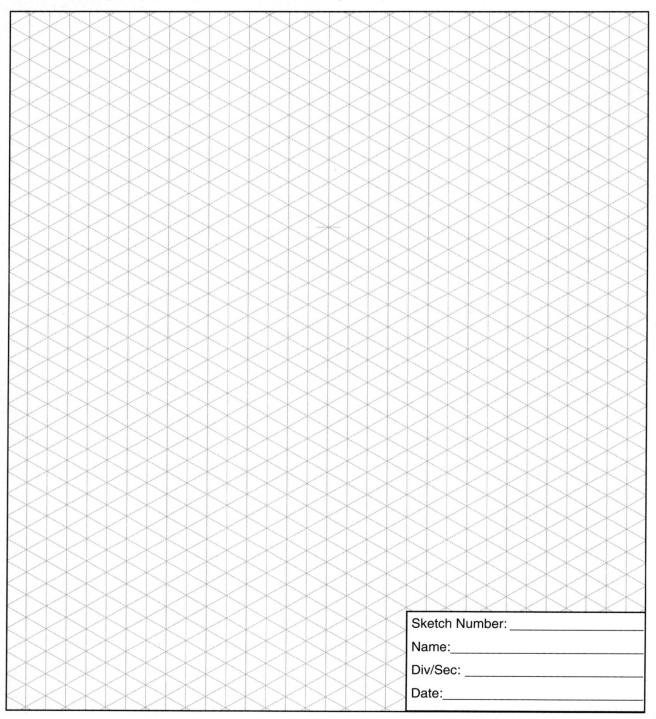

Sketch Number: _____

Name:_____

Div/Sec: _____

Date:_____

Design Problems

GENERAL INSTRUCTIONS

The following design problems are intended to challenge you to be creative, individually or in a group. Those problems designed specifically for a group are so labeled. The design problems are not meant to teach the design process, as much as how to represent ideas graphically, using drawings and computer models. Any design problem labeled a **concept** means that all details of the solution are not necessary. For example, Problem 2 is a concept automobile highway bridge. Obviously, a highway bridge cannot be completely designed in 10, or even 16, weeks by students. However, the basic concept for a highway bridge can be developed and graphically represented in that timeframe.

For each project, you must create the engineering drawings and models necessary to communicate your solution to others. The engineering drawings should include the following:

1. Initial design sketches.
2. Multiview drawings, with dimensions. (Concept designs only need critical dimensions and details, where necessary.)
3. Sectioned assembly drawings, with parts list.
4. Pictorial drawings of the final design, with parts list where appropriate.

Group projects use a team approach to solve the design problems. The team should consist of four to eight students, randomly selected. Each team should have a group leader to schedule meetings, assign tasks and deadlines, and make sure the team works together as a group to solve the problem.

PROBLEMS

1. **Concept solar-powered vehicle. (Group)** Design a solar-powered concept vehicle, for one passenger, that can travel up to 20 MPH.

2. **Concept automobile highway bridge. (Group)** Design a bridge structure that will carry four lanes of traffic and cross a river. The bridge should be 125 feet across and 78 feet above the water in the center.

3. **Concept multipurpose outdoor stadium. (Group)** Design a multipurpose outdoor stadium for football, baseball, and soccer that will seat 54,000 people.

4. **Concept multipurpose indoor stadium. (Group)** Design a multipurpose indoor stadium for basketball, hockey, and volleyball that will seat 18,000 people.

5. **Concept olympic-sized swimming pool. (Group)** Design an olympic-sized swimming facility that will seat 3,500 people.

6. **Ergonomic three-button mouse.** Design an ergonomic computer mouse, with three buttons, that can be used by left- and right-handed people.

7. **Laptop computer case.** Design a hard-plastic carrying case for a notebook computer and mouse that weighs 7 pounds and is 2.25 inches by 8 inches by 11 inches.

8. **Computer disk storage unit.** Design a storage box to hold and organize 100, 3½-inch computer disks.

9. **Computer keyboard tray.** Design a computer keyboard tray that will be fastened to a desk. The keyboard tray must be able to slide under the desk and accommodate a mouse.

10. **Reading lamp.** Design a reading lamp that can be attached to a desk and adjusted to various positions.

11. **Highway interchange. (Group)** Design an interchange between a four-lane highway and a six-lane interstate highway.

12. **Concept airport facility. (Group)** Design an airport facility for a small city to accommodate two airlines and jets no larger than a Boeing 727.

13. **Portable stadium seat.** Design a portable stadium seat, with a backrest that can be folded flat and with storage for a rain poncho.

14. **Cordless telephone.** Design a cordless telephone.

15. **Snow ski carrier.** Design a device that holds four pairs of downhill or cross-country skis and poles and attaches to an automobile.

16. **Computer desk.** Design a desk for a personal computer that uses a 15-inch monitor mounted below the glass top.

17. **Educational toy.** Design an educational toy or game for children ages 3 to 5.

18. **Beverage cooler.** Design a cooler that can accommodate six 12-ounce beverage cans or bottles.

19. **Chair storage device.** Design a portable storage system for 35 folding steel chairs.

20. **Concept railroad bridge. (Group)** Design a railroad bridge that spans a river 100 feet across. It should be 38 feet above the water at the center.

21. **Children's playground. (Group)** Design a neighborhood playground with tennis courts, a basketball court, and playground equipment.

22. **Football/track facility. (Group)** Design a high school football/track facility with concessions and seating for 2,500 people.

23. **Packaging for a computer monitor.** Design the packaging for a 15-inch or 17-inch computer monitor.

24. **Solar water heater. (Group)** Design a solar collector to heat water for a 2,500-square-foot house.

25. **Solar collector for heat. (Group)** Design a solar collector to heat a 1,200-square-foot house located in northern Michigan, at latitude 45 degrees.

26. **Stereo speaker cabinet.** Design compact, three-way stereo speakers for a 40-watt amplifier.

27. **Concept swept wing aircraft. (Group)** Design a swept wing combat aircraft.

28. **Concept commercial aircraft. (Group)** Design a high-speed 350-passenger commercial aircraft.

29. **Concept spacecraft. (Group)** Design a three-person space vehicle to be used by a space station for in-space satellite repair missions.

30. **Remote control.** Design a hand-held infrared remote control for an entertainment center that has cable TV, a videocassette player/recorder, and a stereo with cassette tape, tuner, and audio CD player.

31. **Concept mountain bike. (Group)** Design a lightweight frame for a 26-inch mountain bike with a shock absorbing front fork.

32. **Concept amusement park ride. (G   ) Design** an amusement ride for a specific age   up for a theme park.

33. **Weightlifting bench.** Design a we ifting bench that uses a maximum of 600 pound of free weights. The bench should include the following stations: bench press, dual action leg lift, and adjustable inclined bench.

34. **Stair-stepper exerciser.** Design an exercise stair-stepper with dual hydraulic cylinders that can be adjusted for variable resistance. The unit must have a height adjustment for the handlebar and must fold flat for storage.

35. **Portable basketball hoop.** Design a portable basketball system with a water-filled base that provides 250 pounds of weight for stability. The pole should be adjustable every 6 inches between 7 feet and 10 feet, and the unit must have wheels for easy movement.

36. **Computer workstation desk.** Design a computer workstation desk for a PC-based CAD system. The workstation must accommodate a 17-inch monitor, desktop computer, keyboard, 18-inch digitizing tablet, and some storage.

37. **Concept sports car. (Group)** Design a sports car for two passengers.

38. **Concept electric car. (Group)** Design an electric car to be used by commuters in large metropolitan areas. The car should have a maximum speed of 60 MPH and a range of 120 miles.

39. **Communications device.** Design a communications device that uses pictures, for use by a disabled child. For example, a drawing of a cup with the word "drink" could be used to communicate that the child is thirsty. The communications device must have a minimum of 25 words for basic living functions, such as eat, sleep, open, close, help, etc.

40. **Propane gas grill.** Design an outdoor propane gas grill with approximately 600 square inches of cooking area.

41. **Alphabet toy.** Design a toddler-age toy used to teach the alphabet.

42. **Parking facility.** Design a parking facility to accommodate 100 automobiles, including five handicapped spaces. Include a single entrance with a booth for collecting parking fees.

43. **Logsplitter.** Design a device to split hardwood logs for home fireplaces. The device should use hydraulics to split the logs.

44. **Aluminum can crusher.** Design a device that will automatically crush aluminum beverage cans for recycling. The device should have a feature that automatically loads, crushes, and discharges the cans.

45. **Sports car dashboard.** Design a sports car dashboard. Lay out all the standard features on the dashboard so that they are within reach or eyesight of the driver.

46. **Toothpaste dispenser.** Design a device that will automatically dispense toothpaste from a standard toothpaste tube. This device should be able to mount on a wall.

47. **Ball return.** Design a device that returns a basketball when a person is practicing free throws. This device must attach to a standard basketball rim, backboard, or pole.

48. **File cabinet wheels.** Design a wheeled device that attaches to a file cabinet. This device would make the file cabinet easier to move. The device should be adjustable so that it will fit on different-sized cabinets.

49. **TV/VCR remote control.** Design the controls for a TV/VCR remote control. Take into account the size of an average human finger and its reach. Think of the functions that would be needed and how they could be grouped on the controls. Consider what the overall size and shape should be and what kind of a display it should have, if any.

50. **Door latch/lock.** Design a door latching/locking system. Define the environment (office, home, industrial, etc.) and the kind of door with which it will be used. Take into consideration the characteristics and abilities of a wide range of people who might be using the door. Include such factors as height, strength, and physical handicap.

REVERSE ENGINEERING PROBLEMS

Reverse engineering is a process of taking existing products, evaluating and measuring them, and then creating the CAD database to reproduce them. The following problems can be used as reverse engineering projects. Use a micrometer, scale, and calipers to make measurements. Use manufacturer's catalogs to specify standard parts, such as bearings and fasteners. For each project, the following is required:

1. Disassemble, measure, and sketch each part.
2. Create 3-D models or engineering drawings of each nonstandard part, with dimensions.
3. Specify standard parts, using engineering catalogs.
4. Create an assembly drawing with parts list.
5. Create a written report that summarizes your project, lists the strengths and weaknesses of the product you reverse engineered, comments on the serviceability of the product, and recommends changes to the design, especially as it relates to DFM principles.

The products to be reverse engineered are as follows:

1. ⅜-inch reversible electric hand drill.
2. Staple gun.
3. Electric kitchen hand mixer.
4. Electric can opener.
5. Electric hair dryer.
6. Electric glue gun.
7. Electric hand jigsaw.
8. Dustbuster.
9. Kant-twist clamp.
10. Hold-down clamp.
11. Drillpress vise.
12. Telephone.
13. Computer mouse.
14. Paper stapler.
15. Paper feeder tray for a laser printer.
16. Multiple pen holder for a plotter.
17. Computer game joystick.
18. Piston and connecting rod for an automobile engine.

PROBLEM-SOLVING ACTIVITIES

The following problems can be used as individual or group activities. The activities involve the problem-solving process. Some of the activities also involve visualization ability. Most of the problems will require sketches to solve the problem and to communicate the solution.

1. Determine one solid object that will pass completely through the circular, triangular, and square holes. (Figure 1) The object must pass through each hole one at a time and be tight enough that little or no light passes between it and the sides of the hole. Make an isometric sketch of your solution.
2. Sketch the missing right side view and an isometric view of the two given views in Figure 2. The solution cannot have warped surfaces.
3. Create a low-cost lightweight container that will prevent an egg from breaking when dropped from a third-story window onto concrete.
4. Create a method of filling a 2-liter plastic soda bottle resting on the ground, from a third-story window.
5. A mountain climber starting at sunrise takes a well-worn path from the base of a mountain to its top and completes the trip in one day. The mountain climber camps overnight on the mountain top. At sunrise the climber descends the mountain along the same path. Even though the rates of ascent and descent are different, there is one point along the path which the mountain climber passes at the same time of the day. Prove that there is a single point along the path where this occurs, and make a sketch of your solution.
6. Build the longest cantilevered structure possible using 20 sticks of spaghetti and 24 inches of clear tape. The base of the cantilever must be taped to an 8-inch-square horizontal area. The cantilever must be constructed in 30 minutes or less. When finished, measure the length of the cantilever section from the point on the base nearest to the overhanging cantilever end to the end of the cantilevered section.

Problems 5 and 6 are adapted from *Experiences in Visual Thinking*, R. H. McKim, 1972.

Figure 1   Problem-Solving Activity 1
Adapted from *Experiences in Visual Thinking*, R. H. McKim, 1972.

Figure 2   Problem-Solving Activity 2

WHERE TO FIND CASES

Many engineering case studies, histories, and problems have already been developed and can be used to design your own cases for little or no cost. Here are some sources:

The Center for Case Studies in Engineering

Campus Box 139
Rose-Hulman Institute of Technology
5500 Wabash Avenue
Terre Haute, IN 47803–3999

The catalogue of ASEE engineering cases can be accessed free of charge through the Internet; there is a small per-page fee to order copies of the cases.

http://www.civeng.carleton.ca/ECL/

National Engineering Education Delivery System (NEEDS)

Supported by the NSF-funded Synthesis coalition—including the University of California at Berkeley, Southern University, Stanford University, Cornell University, and Tuskegee University—NEEDS makes course materials, including cases, available through World Wide Web and telnet. It can be accessed through:

http://www.needs.org

The University of Michigan

UM has developed several interactive case study modules that are available from:

CACHE Corporation
Department of Chemical Engineering
University of Texas at Austin
Austin, TX 78712
CACHE@utxbm.cc.utexas.edu

Other sources of cases

H. Scott Fogler, *Strategies for Creative Problem Solving*. Englewood Cliffs, NJ: Prentice Hall, 1994.

Henry Petroski, *Design Paradigms: Case Histories of Error and Judgment in Engineering*. New York: Cambridge University Press, 1994.

Industry Application

The Design of the Boeing 777 Commercial Aircraft*

Computers Aid the Design and Manufacture of the Boeing 777

In Boeing Commercial Airplane Group's 777 Division, computers are modifying several processes, both in Engineering and Operations. In Engineering, traditional drafting boards are becoming a tool of the past as the 777 emerges as the first airplane to be 100 percent digitally designed and "pre-assembled" on computer screen. In Operations, the pencil-and-paper tracking system formerly used to log an airplane's assembly status is being replaced by a more efficient computer system.

In designing the airplane, however, the change is more far-reaching than a simple move from 2-D drawings to 3-D computer images. Boeing officials expect a significant impact on the entire process of designing and building an airplane—improved quality, reduced costs, fewer changes and errors, and less rework. Typically, change, error and rework are the highest cost factors for any manufacturing endeavor.

Computer-Aided Design

Boeing first began using CATIA (Computer-Aided, Three-Dimensional Interactive Application) in 1987 to design some of the parts for the 747-400. Boeing increased its CATIA experience in the following years, with the company augmenting the software to meet its unique requirements. This 3-D software was developed by Dassault Systemes of France and is marketed in the United States through IBM.

Beginning with the 777's launch in October 1990, CATIA use has increased, with the 777 Division now using approximately 1700 CATIA workstations in the Puget Sound area to finish the initial design of the new twinjet. These workstations are supported by four connected IBM mainframe computers, which represent the largest mainframe installation of its kind in the world. This mainframe cluster also is linked to mainframes and workstation installations in Japan; Wichita, Kansas; Philadelphia; and other locations. At the peak of the design process, the 777 Division used more than 2,200 CATIA workstations in the Puget Sound area attached to an eight-mainframe computing cluster.

CATIA is a CAD/CAM (computer-aided design/computer-aided manufacturing) software system that enables engineers to design parts for the 777 as 3-D, solid images and then simulate the assembly of those parts on the computer. The ability to "pre-assemble" parts electron-

Figure 3   Computer Designs Make Costly Interferences Avoidable

Bright colors and sophisticated shapes make the Boeing 777's forward fuselage look more like a work of modern art than an engineering design. Using sophisticated digital computer technology, 777 designers are able to create designs such as these 3-D solid images rather than 2-D drawings. For example, in this picture, the green ribs represent the airplane's structure and the red pipes are part of the flight control system. Designers can pre-assemble, or mock up, portions of the airplane on the computer screen to detect misalignments and other interference problems. This capability will greatly improve accuracy, reduce the need for full-scale mock-ups, and enhance the overall design process.

ically on the computer screen allows engineers to find, and easily correct, misalignments and other fit or interference problems. It also eliminates the need for building a costly full-scale mock-up. (Figure 3)

The application also is an interactive communication tool that allows the various disciplines to work on the airplane's design concurrently rather than sequentially. In the

(continued)

*Courtesy of the Boeing Company.

past, analysts and engineers who worked with 2-D drawings took turns adding elements to the design. Completed designs then were released to production specialists who developed tools and manufacturing plans but rarely were involved at the initial design stage.

Because it was nearly impossible to verify fit and check for other interferences on 2-D drawings, it was essential to build full-scale mock-ups of the airplane being designed. When the inevitable discrepancies were revealed at this stage, hours of costly redesign and retooling were required to correct the problems. In addition, without manufacturing involvement in the initial design, parts were sometimes unnecessarily complex, making them difficult and expensive to produce.

People, Processes

Today, however, on the 777 program, design/build teams—with members from all the participating disciplines—are involved from the beginning of the design. Employees share knowledge and identify problems before tools or parts are ever built. At the peak of the design process, 238 teams of employees from diverse disciplines worked on the 777 design. Subcontractors also join with Boeing design/build teams to ensure that the parts they supply are compatible, and some teams include representatives from 777 airline customers.

The benefits of the team approach are many:

- Improvements in the quality of designs, resulting in dramatic reductions of engineering change orders in early production.
- More reliable and efficient processes because design is done concurrently, not sequentially.
- Reduced manufacturing costs because all involved disciplines have integrated the parts prior to production.
- Decreased scrap and rework.

Now, engineers can view shaded designs as multicolored solids on the workstation display screens. These digitally defined designs are far more accurate than 2-D manual drawings.

Each airplane component can be studied from any perspective, and cross-section views can be easily extracted from the solids. Engineers also use CATIA to accurately analyze weights, balance, and stress on parts and systems.

Figure 4 **Cab of First 777 Assembled by Boeing**

In June 1993, Boeing mechanics completed assembly of the cab of the first 777 twinjet at the company's Wichita Division in Kansas. The cab was shipped by rail in a custom-built car to the final assembly facility in Everett, Washington. In addition to the airplane's nose section, Boeing Wichita has design and manufacturing responsibility for the airplane's seat tracks, engine struts, and nacelles.

The computer-design system will continue to benefit customers of the 777 long after the initial design is complete. As airlines order 777s, the designs are being accurately revised to accommodate individual customer-requested options, such as seat configurations and the location of galleys and lavatories. Additionally, the 777 will grow into a family of airplanes, and future models will benefit from the precise, exacting digital design data of the initial models.

Computer-Aided Manufacturing

In addition to designing the airplane with computers, 777 employees are using computers to make the manufacturing process run more smoothly as well. The 777 Division is using a new system called Assembly & Installation Shop Floor Control, designed to make assembly-job status readily available to shop employees, managers and support organizations. (Figure 4)

(continued)

Industry Application

The Design of the Boeing 777 Commercial Aircraft—*Concluded*

Figure 5   **First Boeing 777**

Functional testing and installation of flight test equipment continues on the first production 777 jetliner, in the foreground. Behind it, employees join together body sections of airplane No. 2, which was used solely for static testing.

Figure 6   **Boeing 777 Scale Model**

This highly detailed, 40th scale model of the Boeing 777 provides a preview of the new airplane. Boeing Commercial Airplane Group has made significant progress on the newest widebody, including essentially completing the airplane's design, beginning major assembly, nearing completion of the expansion at Everett, Washington, to support the airplane's production, and continuing an extensive testing program to ensure a "service-ready" airplane that performs as expected.

Departing from the standard pencil-and-paper method of tracking progress on each airplane within their area, shop employees now use the computer to maintain an instantaneous pulse on the manufacturing process. Workers use technology similar to grocery store bar-code scanners to record status codes, work packages, and employee names when logging manufacturing activity status. The process takes about 20 seconds and provides a real-time snapshot of current activity—managers don't have to wait for the data to be tabulated to know exactly how the airplane is progressing. (Figure 5)

One of the principal goals of the 777 program is to demonstrate an improved process for designing and building airplanes—a process that improves the quality of the end product and dramatically reduces design changes and errors, and the resulting rework. Whether in designing or building the airplane, the use of computing technology is playing a significant role in helping Boeing employees reach their goals of delivering service-ready airplanes on schedule and providing better value to customers. (Figure 6) ■

Additional Problems
and Worksheets

Problems

7.1 Create a single-line drawing of the process flow diagram shown in Figure 7.1.

7.2 Create a single-line drawing of the mechanical flow diagram shown in Figure 7.2.

7.3 (Figure 7.3) Redraw the block diagram sketch as a complete production drawing.

7.4 (Figure 7.4) Redraw the cabling sketch as a complete production drawing.

7.5 Use a CAD software program to make a complete set of production drawings for the electronic circuit shown in Figure 7.5.

Figure 7.1   Process Flow Diagram

(From *Process Piping Drafting,* by Rip Weaver. Copyright © 1986 by Gulf Publishing Company, Houston, TX. Used with permisiion. All rights reserved.)

Figure 7.2   Mechanical Flow Diagram

(From *Process Piping Drafting*, by Rip Weaver. Copyright © 1986 by Gulf Publishing Company, Houston, TX. Used with permission. All rights reserved.)

Figure 7.3   Block Diagram Sketch

Figure 7.4   Cabling Sketch

Figure 7.5   Electronic Circuit

7.6   (Figure 7.6) Draw the highway interchange, using the given dimensions.

7.7   Model the production facility shown in Figure 7.7. Before beginning the modeling process on the computer, divide the facility into a series of logical parts, giving each one a name. You may find it helpful to create sketches of the individual parts.

7.8   (Figure 7.8) Many power plants are installing large hyperbolic natural-draft cooling towers similar to that shown. They are thin, light-weight concrete, often no more than 4″ thick. An open grid of beams at the bottom supports the shell and permits the entry of air. Draw or sketch the cooling tower.

7.9   (Figure 7.9) An industrial building has a roof supported   by split-ring connector timber trusses as shown. (The 4″ × 6″ struts all lie centered on the truss center lines.) Draw or sketch the roof truss.

Figure 7.6   Highway Interchange

Figure 7.7   Create a 3-D Model of this Production Faciliy

Support cables

Gore dia 2.24´

520´

370´

Support cables

Diffusion grid

160´

80´

380´

Figure 7.8 Hyperbolic Cooling Towers

(Courtesy of the Marley Company, Kansas City, Mo.)

Figure 7.9 Timber Truss

7.10 Draw the three logic gates shown in Figure 7.10. Assume symmetry about a central horizontal axis.

7.11 Draw the open traverse shown in Figure 7.11.

7.12 Complete the drawing of the column and base plate in Figure 7.12. Estimate all dimensions which are not given.

7.13 Complete the drawing of the retaining wall shown in Figure 7.13.

Figure 7.10   Logic Gates

Figure 7.11   Open Traverse

Figure 7.12   Steel Connection Detail

Figure 7.13   Retaining Wall Section

7.14 Make a drawing of the piping layout shown in Figure 7.14.

7.15 The slide gate shown will hold a $\frac{1}{2}''$ steel plate to give access to a tank. Complete the oblique and orthographic drawings shown in Figure 7.15.

7.16 The isometric drawing in Figure 7.16 shows the method for securing $1\frac{1}{2}''$ steel grating to a 2 × 2 angle bracket. Complete the drawing as shown estimating any dimensions that are not given.

7.17 An engineer wishes to locate a house on a plot of land with a 15° angle from due north. Draw the house per the dimensions shown in Figure 7.17 and then *rotate* it to place on the plot plan..

7.18 Complete the drawing of PCB1 shown in Figure 7.18. Change the sizes of the holes to create PCB2.

Figure 7.14   Piping Isometric

Figure 7.15   Slide Gate

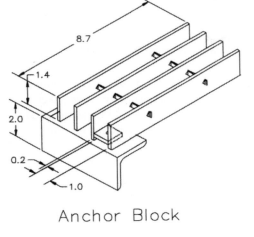

Anchor Block

Figure 7.16 **Anchor Block**

Figure 7.17   **Plot Plan**

PCB1   PCB2

Figure 7.18   **Printed Circuit Boards**

7.19 Draw the W10 × 15 beam shown in Figure 7.19.

7.20 (Figure 7.20) An architectural drawing using ***Architectural Units.*** The inside line is 5″ away from the outer.

7.21 (Figure 7.21) Given the two views of a multiview drawing of an object, sketch or draw the given views or use CAD, and then add the missing view. As an additional exercise, create a pictorial sketch of the object.

7.22 (Figure 7.22) Given the three incomplete views of a multiview drawing of an object, sketch or draw the given views or use CAD, and then add the missing line or lines. As an additional exercise, create a pictorial sketch of the object.

W10X15

Figure 7.19   I-Beams

Figure 7.20   Floor Plan

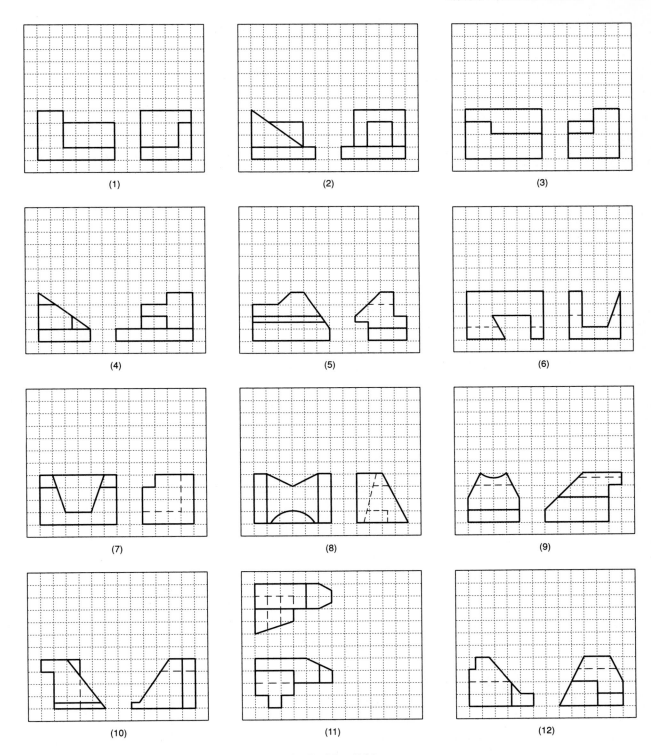

Figure 7.21   Two-View Drawings of Several Objects, for Problem 7.21

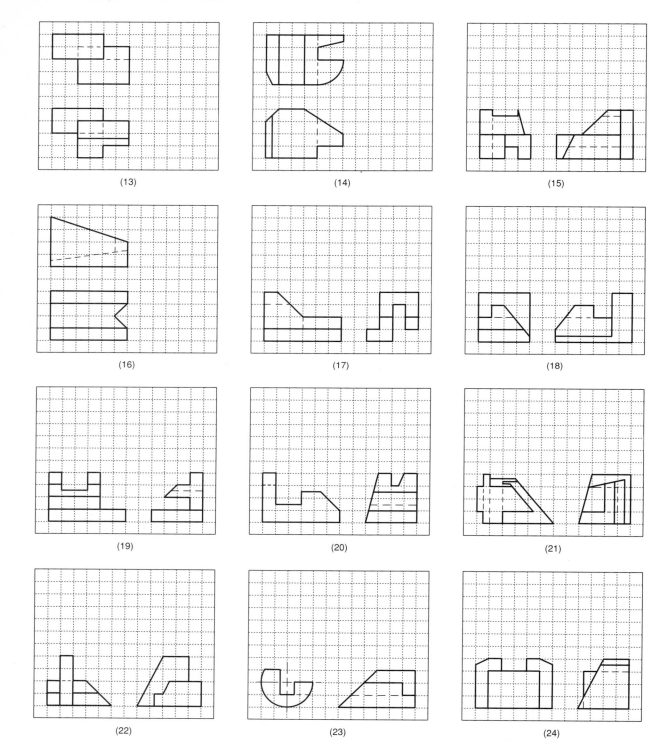

(13)

(14)

(15)

(16)

(17)

(18)

(19)

(20)

(21)

(22)

(23)

(24)

Figure 7.21 Continued

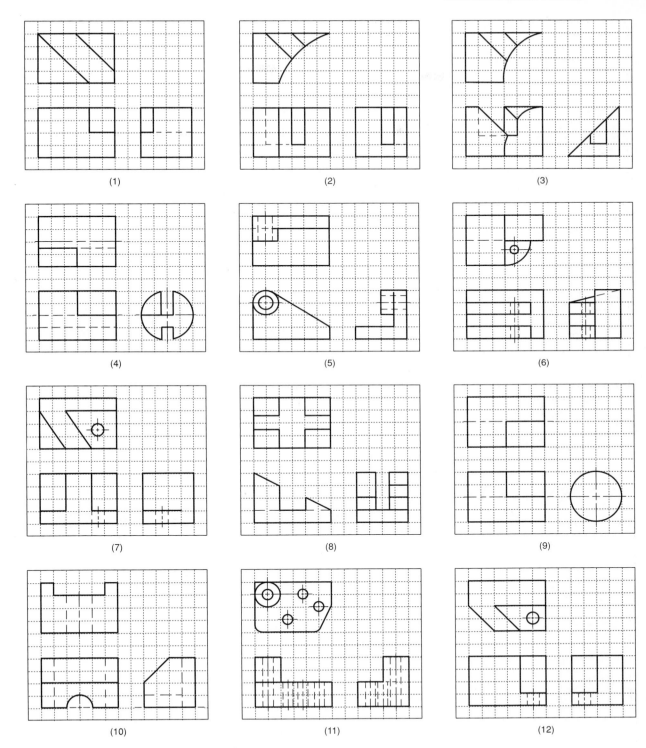

Figure 7.22   Three Incomplete Views of a Multiview Drawing of an Object, for Problem 7.22

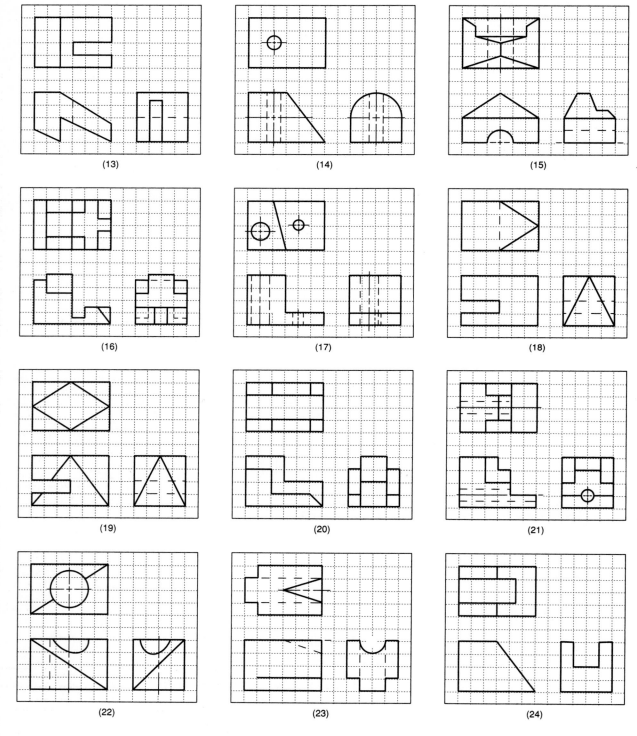

(13)   (14)   (15)

(16)   (17)   (18)

(19)   (20)   (21)

(22)   (23)   (24)

Figure 7.22 Continued

Orthographic Sketch Paper

Sketch Number: _____

Name: _____

Div/Sec: _____

Date: _____

Orthographic Sketch Paper

Sketch Number: _____

Name:_____

Div/Sec: _____

Date:_____

Orthographic Sketch Paper

Sketch Number: _____

Name:_____

Div/Sec: _____

Date:_____

Orthographic Sketch Paper

Sketch Number: _____

Name:_____

Div/Sec: _____

Date:_____

Orthographic Sketch Paper

Sketch Number: _____

Name:_____

Div/Sec: _____

Date:_____

Isometric Sketch Paper

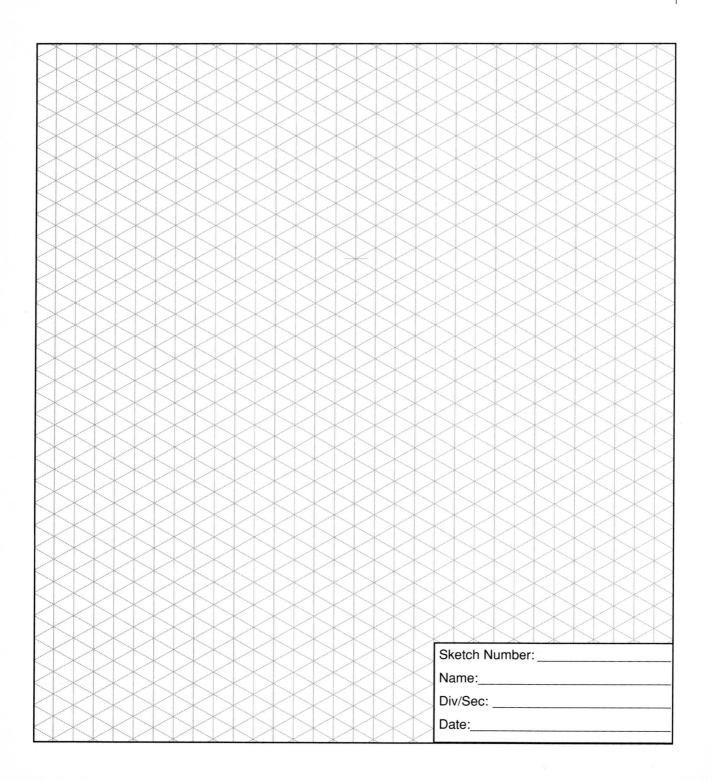

Sketch Number: _____

Name:_____

Div/Sec: _____

Date:_____

Isometric Sketch Paper

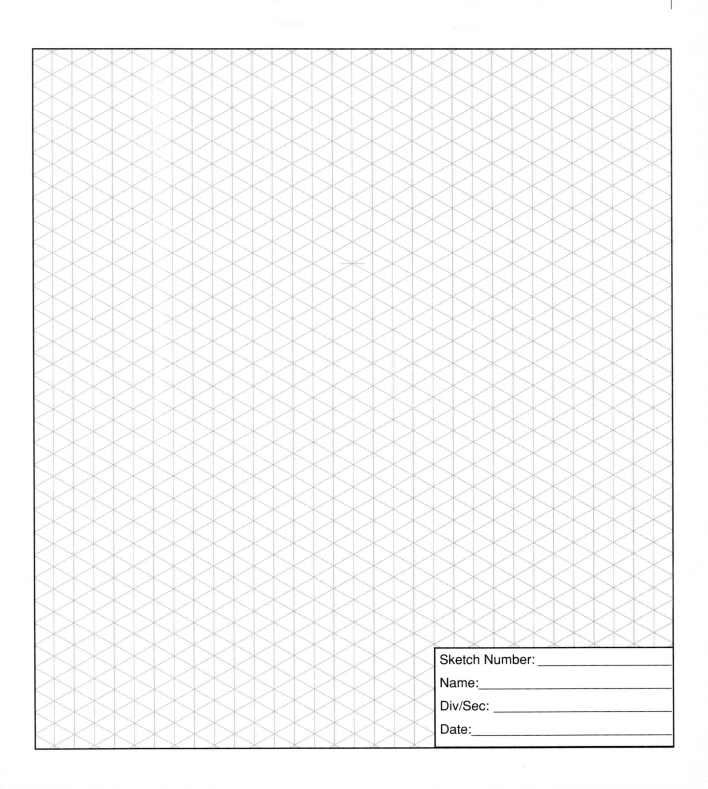

Sketch Number: _____

Name:_____

Div/Sec: _____

Date:_____

Isometric Sketch Paper

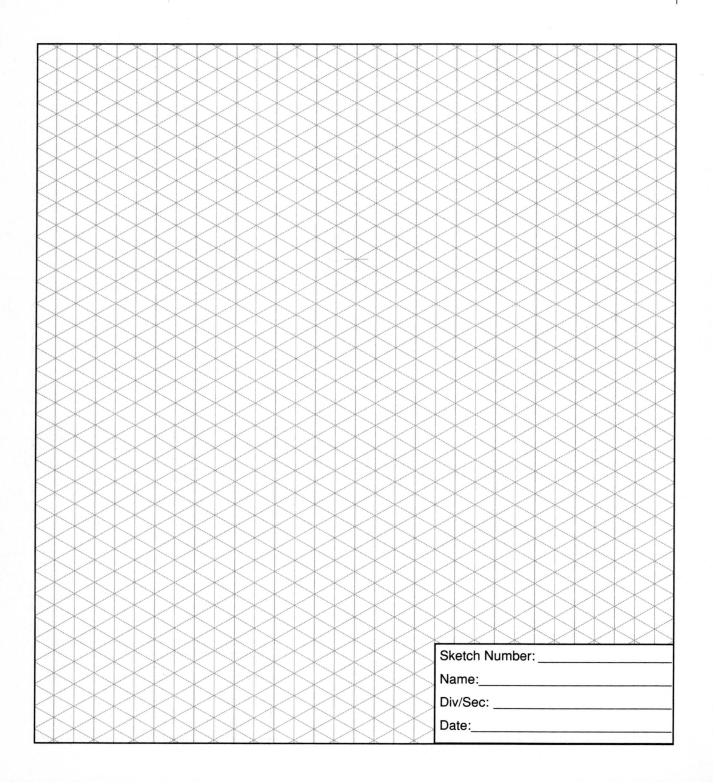

Sketch Number: _____

Name:_____

Div/Sec: _____

Date:_____

Isometric Sketch Paper

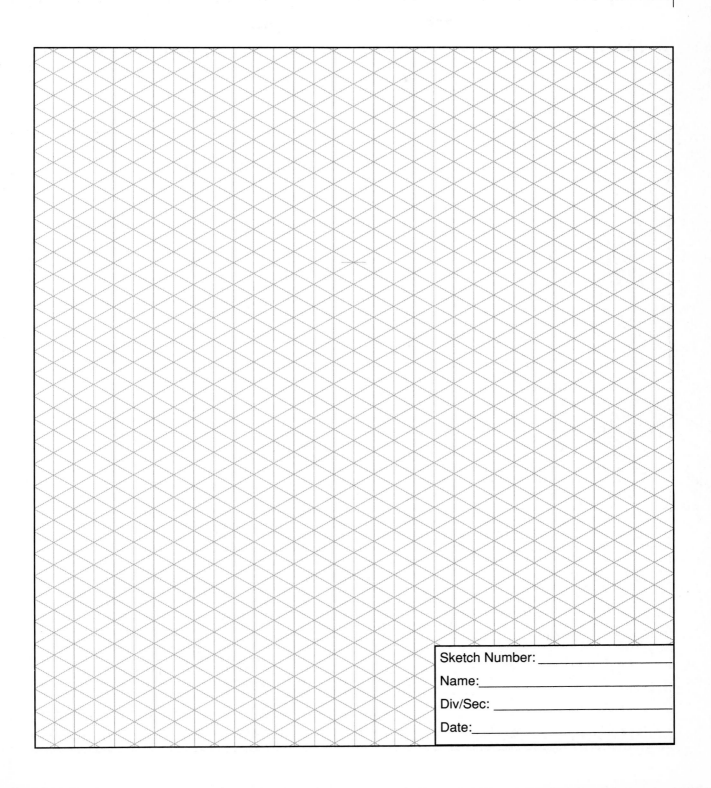

Sketch Number: _____

Name:_____

Div/Sec: _____

Date:_____

Isometric Sketch Paper

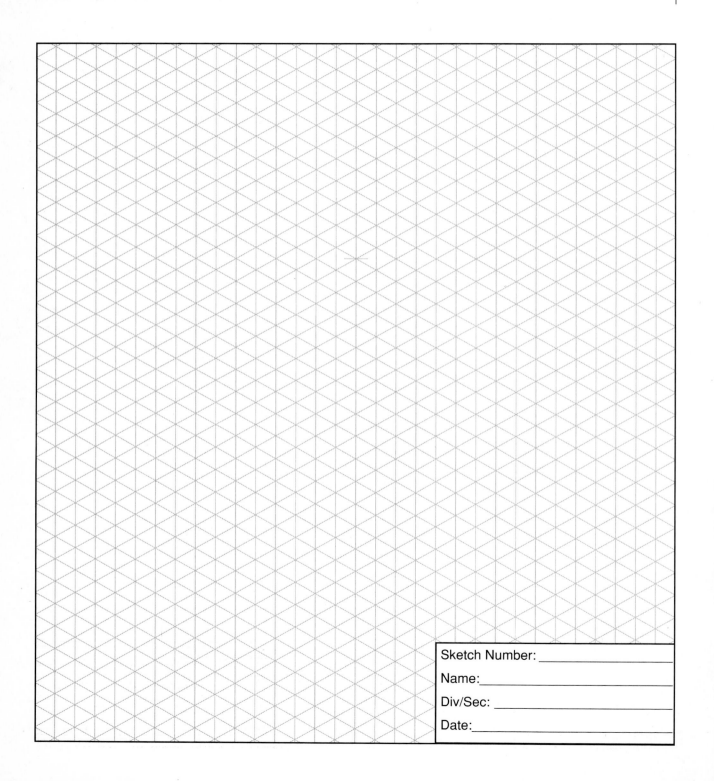

Sketch Number: _____

Name:_____

Div/Sec: _____

Date:_____

Problem Worksheet

Sketch Number: _____

Name: _____

Div/Sec: _____

Date: _____

Problem Worksheet

Sketch Number: _____

Name:_____

Div/Sec: _____

Date:_____

Problem Worksheet

Sketch Number: _____

Name:_____

Div/Sec: _____

Date:_____

Problem Worksheet

Sketch Number: _____

Name:_____

Div/Sec: _____

Date:_____

Index